SOLVING IT SERVICE MANAGEMENT

Antonio Narro

Wholesale discounts for book orders are available
through Ingram Distributors.

Tellwell Talent
www.tellwell.ca

ISBN
978-0-2288-1346-0 (Hardcover)
978-0-2288-1345-3 (Paperback)
978-0-2288-1347-7 (eBook)

TO MY LOVELY WIFE GLORIA

INTRODUCTION

Today's IT environments are more complex than ever. Hardware equipment is all over the place, you have all kinds of application teams working on your environment, even from competitive companies, and your hardware technicians are literally sitting down at the other side of the planet. Besides that, you have all kind of procedures and methodologies to follow. The good old times when you just rebooted a server by telling a couple of colleagues about is now just a nostalgic souvenir. You now need to complete complicated paperwork and go through all kind of bureaucracy to get things done.

Besides, everybody's purpose in business seem to be centred in cutting costs. All budgets are now dramatically reduced and you don't have the luxury of having enough resources to complete the job. We now have to deal with people working together, but living in different time zones and with totally different culture. It sometimes feels like directors are betting on you and playing games to see if you can still solve all kind of problems even without the most essential tools. It feels like if they're joking in the background and saying: "now let's take away his local hardware support, or maybe if we shut down his internet, what if we give him an IT administrator that doesn't evens speak his language?" What a circus!

Despite all of this, there is still hope. There are plenty of ways to get around and accomplish a good professional job, provide an excellent IT service and have happy customers. We are sometimes so busy looking at the imminent issues that we often forget to step out, take a look at the big picture and do what really makes sense. I'll provide some personal stories about this, I actually laugh about them now and I hope it will at least make you smile. But more importantly, I hope it helps you better deal with the current issues in the IT environment, follow the proper procedures and avoid falling in ridiculous behaviour that is so common in today's IT environment.

This book will not only help IT managers, but it's also made in a way that business directors, VPs and managers will better understand the IT environment, even without having much IT knowledge. It is made on an easy to read format that most people on the business will understand. You may want to refer to the table of contents if you want to check on a special section.

I'll first talk about the common mistakes while applying today's rigorous IT procedures. When something doesn't make sense, it's probably because it's not right. People keep on doing things and they justify their behaviour by saying they are only following the procedure. Some other people may argue that things have always been done in a certain way. However, this needs to be reviewed.

I'll then talk about two subjects that will help you better control your IT environment: Metrics and Data Safety systems. If you don't have much knowledge about IT, these sections will help you better understand the importance of certain systems and ways of working in IT.

The core of IT management is incident, problem and change management. These subjects are treated in detail in the next 3 chapters. We will discuss in detail how to apply procedures in a way that will actually help you solve the issues quicker and more efficiently. I'll finally talk about the Service Desk. It's sad to witness how this important department has been neglected in recent years. This last chapter demonstrates how assisting the end users can be done in a useful and professional way.

I hope you enjoy your reading!

CHAPTER 1

Strategic procedures

There's nothing better that coming in early in the morning, pick up a warm cup of coffee and get ready to start your day at the office. That is of course, until the day really starts...

Oh, the pager just went off! What is this? What is going on? Hmmm... let me call in to the bridge. After dialing a bunch of numbers, hearing greeting in two official languages and reminders about muting 1 finally get in. "Hi this is Tony. What is the situation?" Now I have to wait for the answer from the incident manager who is trying to coordinate the emergency. There are lots of people talking to each other on the bridge. Eventually somebody will even ask how's the weather at the other side of the planet.

Now I finally get an answer: "This is Bob from incident management, application X is down, we suspect a hardware problem. We have Suzanne from application support, Joe from the system supervision, Han and Ho from the Unix support team and Miguel from the Database support team." Wow, quite a big gang on the line.

Now, let's get to work. As an infrastructure support manager, I checked on my Unix support team guys and see how they're doing. The poor guys are struggling between answering questions and doing their investigation. Even worse, one of them doesn't have a headphone so he's just using his speaker phone while he's working. "Han, go on mute!". Wow, finally some silence, until somebody asks again: "Han, how is the server doing?". Silence breaks again and it's just like if we placed a mic in the middle of downtown Delhi. You can hear lots of people talking in the background, some traffic noises and even a dog barking from time to time. My mind tries to apply Dolby filters as much as it can and concentrate on what the technician is saying. Yup, finally it looks like one of the disks has failed.

Next logical question: "How come the disk failed?" to what follows the traditional answer "It happens". OK, let's call hardware support now, and yes, time for a second coffee!

After a few minutes which seem to last for the rest of the morning we finally got a beep: "This is Yale from hardware support, what can I do for you?". Following to that, the incident manager re-explains the situation for the 10th time and gives the list of people on the bridge, rightly following his incident management procedure.

Wow, wait a minute. I missed a few bips on the bridge. How many people are actually on the bridge? We had the incident manager, the application support technician, the lady responsible for application support, the application owner, the database manager, 2 Unix support system administrators, the infrastructure support manager, the hardware support technician and the Director Executive on the bridge.

PROCEDURE ANALYSIS

We have nine people on the bridge for a stupid hard disk failure! All these people are wasting at least 2 hours on this incident and they are all getting paid for it. How is this cost efficient? And we're still not counting the people who are following this incident in the background, people supporting the applications used for managing the incident, plus all the people who will eventually be involved in incident reporting and follow-up. Everybody is following the right procedure. Nevertheless, this doesn't seem to be right. We have way too many people working to solve a trivial problem, so it's important to take some time off, step back and analyze the procedures being used to understand if this is the best way to manage IT resources.

Every company is looking on how to cut costs, but IT support expenses have sky rocketed in the last decade. We now have more efficient servers, hard disks and network equipment. We are even willing to pay more for them because we look at it as a good investment, just as we do when we buy a car. If we can afford a more reliable car, we'll go for it knowing that we'll save on troubles and repair costs. However good IT equipment is just part of the puzzle.

We need to verify how the equipment is being used and how our IT management procedures align with our business. Failing to do so will normally cause incident headaches like the one described above. Having more people looking at an issue will not necessarily speed up recovery nor will it help improve the long-term service.

To illustrate and better explain this point, we can make the comparison between our IT support teams working together to solve an issue and a team of people digging a hole in the ground by hand in a construction field. How many people can work simultaneously on this job? Well, one should normally be enough. Maybe even two if the hole is big enough. Having more people working on it will just make things complicated as the diggers will stumble and block each other while doing their job.

Now, how do you make sure the job is done right? Will you have a bunch of supervisors with procedures and plans reading them to the diggers and asking for status every five minutes? How about having the cement team standing by ready to work once the digging job is ready. Of course not, every reasonable person will normally understand that it will be just a waste of time. We will normally give clear instructions to the working people on how we want the work done. We will also hire skilled workers. People who know and have the experience doing this job, even if it's considered a simple task. We need to have people we can trust and don't need to be guided every minute.

As trivial as it may seem, digging a whole on a construction project is critical. If it's not done right the whole project can be compromised. It hast to be done deep enough and with the right measures, according to specifications. But with the right people, some sporadic supervision and good measurement procedures we never give a second thought to that part of the project. It gets done right, on time and according to specifications.

Now, can we apply the same principles to IT incident management? I agree that a construction project may be different, but we can certainly learn by observation of what works and what is the best way to get the job done right. We can easily understand that having more people working on a certain task will not necessarily improve performance or accelerate the job's completion. We can also understand that micromanaging every task is normally useless and counterproductive.

We must definitely revise our current methods in order to improve the way we work and have a more efficient system for using our resources. Someone may object by saying: "But we're using the most recommended IT procedures in the industry!". IT management has come a long way, since the time of the cowboy IT programmer to the more sophisticated and long approved methods used today. On the other hand, going back to my construction example, will you use somebody else's blueprints to build your house? The answer may be yes, if we're building the exact same house. Otherwise we may pick some useful information from it and adapt it to our own needs.

Widely accepted IT procedures are built to cover every aspect of project, change and incident management. As good and professional as they are, they may miss to cover some important aspects of our very own particular business. Then again, they might be overkilled in certain situations. The biggest mistake is to apply them overall to our whole IT environment regardless of any special needs and challenges of each system and without taking the time to customize them to our own specific needs.

By the way, the application that was down in the introduction was used to generate reports. For the person responsible for these reports it was the end of the world, she was not going to be able to present her reports on time. What she doesn't know is that those reports are no longer considered as important as they were before. Most sales directors who used to depend on them have now turned their attention to other ways to gather the information they need for their projections and only use these reports as a reference. There was way too much attention paid on an IT system that was not company critical application and was not affecting the business. Still, the person responsible for this system had only one thing in her mind: bring the application back as soon as possible.

Other people in the company had other priorities and rightly so. There was a new project being studied on database performance and optimization. Some Unix servers needed to be checked and several of them required immediate patching. Some applications had to be revised. Yet, all of these people were wasting time following an incident management procedure because it was affecting somebody in the company. What a waste of business focus!

When IT incident management was deployed in this company, it was done on the whole environment. Although the application was identified as not being business critical, it still made part of the list of applications that need to follow the cumbersome complicated procedures as all other systems in the company.

This shows the importance of revising the way we manage the IT environment. Are we following certain procedures just because Joe said so? Because everybody says that's the way to go? We should take some time to analyze he IT business needs in our company and see how can we better apply the right procedures at the right time for the right system.

We can find an excellent example of this in the airline industry, which is widely known by its strict application of procedures. Before takeoff, pilots must go through a detailed checklist every time. This procedure was created to ensure security for all passengers and regulate the industry. However as good as this checklist may be, they cannot apply the same list to every airline model out there. They use a specific checklist that has been created and adapted to each single model with its own special characteristics and properties.

How about creating a universal takeoff procedure that will cover all airplane models? Wouldn't that make things easier? Although it may be technically possible to do that it will certainly be unwise. The result of such a list will be a cumbersome, lengthy, difficult to follow procedure. Takeoff delays could become the norm. Even worse, this increases the margin of error by pilots trying to follow this new, universal and long procedure. They might be tempted to run through some items a little quicker which may end up in oversight or even catastrophe.

Does this sound familiar? IT processes and procedures nowadays have become lengthy and cumbersome. Project managers often complain that their projects are getting delayed due to bureaucratic procedures. Change management is a total new challenge, especially when you need to introduce new important changes on the system. People often complain about the response time on incident management. In addition, the service desk (generally known as the help desk) has become something to avoid unless you want to chat with a total stranger for an hour or so.

These examples clearly illustrate the need to revise the way we apply procedures in today's IT environment. One of the first solutions is to adapt our IT procedures to our different systems in our company instead of creating just one general procedure and applying it globally. Now, I do understand that there may be dozens or even hundreds of different systems used in a company. Trying to create an adapt procedures to each single system will be a nightmare in itself. In addition, every single procedure becomes a living entity and needs to be maintained regularly (We'll cover this subject in a different chapter). Evidently, we cannot just create a different procedure for every existing application or system used in the company.

Nevertheless, in most cases we can group applications in different categories or application families. These systems may be similar in criticality and importance. However, they don't need to share the same audience, clients or business units. They will share the same procedures as we intend to have a similar service from our IT department regardless if it's an internal or external service.

The most common practice is to divide our systems by their technology or OS platform. Although this is quite useful for the technical teams it doesn't really serve the business purpose for the application. Although we must seriously consider the technology used by the application in our procedures, it should not be the determining factor. In some cases, you may have an application running on a variety of different servers using different kinds of technology. It's often quite confusing when the IT department is dividing their IT services by OS platform. Even though servers need to be treated differently according to the technology being used, creating this kind of division will often result in losing the big picture on how we are serving the business through our IT service.

It is true that this method may require having different people attending different parts of the system. However, it will not make a difference if we divide the procedures by technology or by business purpose, especially if we are using the "technology expert role" model often contracted by IT services. One way or the other, we may need people with different sorts of expertise working on our system. On the other hand, there is an advantage in marking the division by business application; all of the people assigned to this system will be following the same procedure and they will all be concentrated in the final goal: to provide a professional IT service to the business. This will avoid different teams playing the blame game (accusing one another saying that their part of the system is OK and it's the other part that is at fault). It will also discourage the habit of technicians walking away once they feel their part of the job is done. They will all be responsible for making sure the system is up and running efficiently regardless of the OS system or technical platform being used.

Nevertheless, this method does require to study all applications in the company. Some people may object and find this tedious, even useless. Some other people may consider it as a waste of time. The reason is that normally when we contract IT services we just need to provide a server inventory and voila! It's done! But let's don't lose the big picture again. We may think we had finished our job once the IT contract or general procedure is signed. We need to be very careful, speedy jobs are not often the best quality jobs. By doing things too fast you may go from hero to zero in no time. Once the IT services start their job you may become unpopular very quickly when people start experiencing the usual problems on regular IT services.

On the other hand, there are a lot of benefits in taking the time to know your environment, understanding your needs and applying the right solutions to the right systems. Evidently, any time taken to better understand your business and your needs is time well invested. Far from being a waste of time and resources it will also help you focalise on the important parts of your business while keeping the whole picture in a good balanced view.

BENEFITS

Regardless on which side of the negotiating table you're sitting on, there is a clear cost benefit in applying strategic procedures for the systems in the company rather than just applying a general cover it all procedure.

If you're on the IT services receiving end you may find useful to know that you don't need to buy a high end superior service for your whole environment. You may be able to actually reduce the number of servers supported under the contract. You may then determine which servers must absolutely be covered and which are optional. While putting a spotlight on your main systems you may ask for second or third degree coverage for the rest of the environment.

On the opposite end, if you're an IT service provider you may have better negotiation arguments with your client. You can actually show you're interested in their business by determining which kind of coverage is better for them. I understand we are always under pressure the make the maximum sale and high server count is normally well rewarded. Everybody needs to cut costs and IT service companies are selling their services are very low prices. But on the long-term, selling bellow profit will hurt your business so it's better to lower prices by using good IT wisdom. Otherwise just enjoy your high stock market fantasy while it lasts knowing that you can't live in a dream world forever.

As much as direct cost is highly important in today's market, efficiency is actually as important if not even more critical to survive in the modern business place. This is probably the more tangible benefit from using targeted procedures instead of general ones. Your business response will be better, you will be perceived as a more professional company and your employees will be happier. Everybody prefers to work in a place where things work and bureaucratic procedures are not stopping good reasonable business practices. In addition, people want to deal with trustworthy companies that are reliable and on which we can consistently count on.

It's difficult to put a price tag on reliability and trust, but its value is unquestionable. You will definitely receive a benefit for working smarter and more efficiently by improving your IT business procedures.

To wrap it up, get to know your company. Take the time to understand how the different applications and systems work together in your business. Identify and value the importance of each of them and then create procedures that better meet your specific business needs.

This book will cover different aspects of IT management like incident management, change management and service desk. You'll need to verify if the IT procedures being used on each category are really helping your environment or not. But just before treating those subjects, let's discuss some IT Management essentials on the next chapters.

CHAPTER 2

Speciality Technicians

Here's another real life scenario about a technician treating a service desk call:

The kind smiley technician arrives to Trouble Joe office and asks: "how can I help you?". Trouble Joe answers back by saying: "My mouse is not moving anymore. The keyboard seems to be OK but I really need to use my mouse and it just doesn't respond!" The technician sits down in front of the computer and moves the mouse to confirm the caller's problem. He then performs what any good technician will do, yes, a re-boot. After the reboot, he finds the computer still has the same problem so he tries to check the machine's configuration. After a few minutes he finally gets his aha moment! "Got it, I know what the problem is, the mouse is disconnected". "Oh wow, such a simple problem" replies Trouble Joe. "Can you please connect it?" he asks.

"Oh no!" answers the technician, "I can't do that! I cannot connect any devices. That's the job of the hardware technician." "Never mind!" cries Trouble Joe, "I'll do it myself". The technician quickly pushes him back and says "You cannot connect your devices yourself or you will void your warranty. If you do that I will report it and we will not be responsible for any hardware failures on your computer".

You may be surprised, but this actually happened. To be honest it happened a few years back, when keyboards and mouse where not using the USB ports. However, it didn't take a University degree to understand how to connect them. It was still quite obvious back then just as it is now. This example shows one of today's biggest problems in IT service management: specialization.

It's a fact, you cannot just call for a general IT technician. If you need some technical help you must be specific, and quite specific. Servers come in all kind of different flavours according to their operating system. They could be Windows or Unix. You then need to subdivide the Unix environment: Is it SUN, HP, Linux, AIX or something else? Of course, Mainframe is a whole different category in itself.

Not only do you need to know if you need a Database guru, but you need to specify if it's Oracle, MS SQL or any other kind of Database. Besides, now you need to specify if you need a DBA or a DBC. Additionally, each application team is totally different from one another. Application team A will have no clue about application B and vice versa. We have subdivided each particular IT discipline and we have a distinct kind of technician for every single one of them.

The outcome of this IT management method is to have scenarios like the one described on the first chapter. The reason for this is that quite rarely will an incident, change or project require just one particular IT discipline. Most activities will involve at least two or more type of technicians. Now, how did we ever get to this?

REASON FOR IT SPECIALITIES
Here's a little history. After what happened with the year 2000 horror scenarios and the ".com" business created an enormous IT boom we understood the importance of managing IT through standardize procedures to help us better control what is going on in the server room. Business people felt like their companies were being hijacked by the IT gurus who believed they knew how to run things. By introducing

rules, procedures and management tactics they finally got their business control back, and IT outsourcing was the way to go.

In order to better manage the IT department, we divided it into different disciplines and skills capacities. Instead of having Joe checking his database, connecting his server to the network and rebooting it when necessary we now have skilled specialists that will make sure that every aspect of the IT service is done in the best possible way. This seemed to be a great idea at first, however the real world is far from this utopic theory. We have divided the IT disciplines so much that we know need a whole team to perform even the more trivial tasks. All of this without even considering the administration, HR and management of all these resources.

Now, let's compare this we what we do in medicine. We're glad to have specialist for every aspect of our bodies. People that really know and concentrate on specific organs instead of going with somebody with just a general knowledge about them. The success in this field has been so great that nobody thinks about ever going back to the old ways and having a general doctor perform surgery as it was done before.

However, general practitioners still exist for a reason, we don't always need a speciality doctor for every illness. People will normally visit a general doctor and only get to a specialist once the doctor find a problem with a specific organ. Yes, some people will skip the first step and go directly to the specialist if they think they know what their problem is. Nevertheless, even this type of patients will sometimes visit a general MD from time to time.

GENERAL TECHNICIAN SOLUTION
OK, now let's get back to the IT field. Where are the general IT technicians? Unfortunately, it seems like they have disappeared. Unless you can somehow find some knowledgeable gurus from the 80's and 90's still around, normally IT technicians are only called by their speciality.

But there's still hope, the system has changed, but the basics are still the same.

The reason for this optimism is that most IT technicians have a general knowledge in other areas. For example, an application programmer will normally know how to start a database and a DBC will know for sure how to reboot a server; so they all do have some knowledge in other areas outside their specialized scope to perform their own work. In reality, the main reason why we use specialty technicians is because that's the way we describe their roles in our procedures.

Here's what's happening: Team A has been contracted to support all servers in the business. Team B is contracted to support the applications. So Team A doesn't want Team B to reboot the servers or to change their configuration because they fear they will lose control of the environment. The consequence of this is that Team A will always call Team B to close the application before rebooting the server. The opposite is also true, whenever Team B is working in an application and needs to reboot the server. The scenario gets even worse if there is a database in the server, we'll then need to call Team C.

Now all three teams may belong to the same company, or they may not. In any case, they need to be responsible for their own piece of the puzzle. Therefore, they will create procedures to limit the work assigned to each team. In the disconnected mouse example discussed earlier in this chapter, there were clear guidelines to protect the hardware team's jobs. So nobody else was allowed to connect or disconnect any devices on their computers.

Somebody may say: "I'm already paying company X to support the servers so let them take care of their responsibilities and reboot their servers!" OK, that may sound reasonable, but to what cost? Are you willing to have your applications down another hour and lose business because you're already paying somebody else for a specific task? It will be dumb to wait for the waiter in a restaurant to put your fork on the right side of the plate and let your meal go cold. Yes, it's definitely their responsibility. Yes, it should have been done, but wouldn't it just be easier to take the fork and enjoy your meal? In today's competitive environment, waiting too long to have IT systems restored can cause your business to go cold.

The other problem with the special technician method is the blame game. "The application is not working because the server has a problem" says Team B. Following by the statement by Team A: "The server is OK; the application is at fault". This may go on for some time while you're there on the bridge waiting for somebody to bring a solution and have the system restored.

The real solution for these kind of situations is to have shared responsibilities. How about having both teams responsible for server's configuration? In other words, making our participant teams responsible for the final outcome, to have a good working system. Yes, it can be done with the proper documentation and communication. Let's use server's configuration as an example. Server's configuration should always be documented anyway. Server support teams should have clearly defined the parameters on their server. The application team normally gives their desired parameters to the server support team so they can apply them, but they can definitely apply the parameters themselves with the proper access. They just need to respect the guidelines defined by the server team and discuss any discrepancies.

On the same point, if the application team is working on their system and require a server reboot they can do that themselves. It's no rocket science! Anybody can do it, even our teenagers at home will often reboot the computer. Yes, I heard the argument: "But what happens if the server doesn't come back?" Well, what happens now? The server support team doesn't have any magical powers. They may need to get to the console or get a physical person in the server room to fix it. It's only a machine, it will not get offended if another team is working on it. Another team may have an early start on troubleshooting and reboot the server. We'll discuss incident management in another chapter, but you can see that basic functions like rebooting a server and changing configuration can be performed by other teams.

But what about responsibility? Won't the server support team complain that somebody else is messing around with their server? Quite frankly, they shouldn't. They are responsible for the server no matter what. Sometimes the server will reboot automatically after a system failure. Or if the server froze because of someone else fault, like if an application used all the available memory, they still need to fix it. Everything should be documented and access should be well determined in order to trace who did what. They can still find out which actions have caused an outage. Having different teams reboot the servers doesn't add any value to the end solution.

Having said that, leave the expert work to the experts. Just as it happens in medicine, there are things outside the basic knowledge that will go beyond what an application programmer or a DBC can do. In that case you may want to call a server specialist to solve those kinds of problems. So you use the specialist for what is meant, special cases, not for general reboots. On a side note, you may actually find out that most OS support system administrators are no real experts. In many cases they just have a general knowledge about the OS and will call the vendor if they run into any problems.

We might not be able to go back to the time where honest Joe was doing all the necessary shores on his server. But we can definitely work better if we extend the team's responsibilities to include just some common tasks using their already acquired basic server knowledge. The same principle can be applied to other fields like different Unix platforms, DBC versus DBA, etc. On my personal experience, most technicians are happy to have additional access and power as this gives them more control on their own systems and they're able to better respond to their client's demands.

CHAPTER 3

Metrics

It is always useful to know how well our services are doing and determine our actions and responses to resolve performance issues. This is the logic behind modern KPIs (Key Performance Indicators), SLAs (Service Level Agreements) and SLOs (Service Level Objectives). All these metrics can help us understand the level of service that we're currently providing.

Nevertheless, we should always be careful with metrics. Have you heard about the guy who drowned in a lake with an average depth of 4 feet? Yeah, that's the average, but the lake is 10 feet deep near the centre were the guy drowned. Metrics can be misleading, or can be manipulated in certain ways when somebody wants to prove an argument.

How about having a server environment with 99.5% reliability? Sounds great, until the server goes down in the middle of the presentation with a client that was going to sign a ten-million-dollar contract. In this case, will you really care that the server was up and running all through the night? Not really, who cares. It's important to have our service respond correctly when we need it. Reliability should be measured when service is really required, otherwise the numbers don't make much sense.

CRITICAL WORKING TIME

Then again, every single system is different and has its own particular needs. Enough of these statements like 8am to 5pm is prime time and the rest is non-prime. It always depends on which system you're talking about. For example: the backup systems usually work at night. Backup servers are normally only used during the day for restores. So, is it really useful to measure the backup system reliability based on prime time? Absolutely not! In this case the servers must be running at night when backups need to be taken.

Let's add another example to show the complexity of the situation. We have a website used to sell vacation packages. How do you determine the prime-time? Well, you should check when is the systems accessed the most. For this example, we may say that 95% of the customer base are people that work from nine to five and normally access the site late during the evening hours. That's the kind of information that will help us determine the real prime time of the system. If the servers are down during the evening, we may lose valuable clients and our sales will be affected, even if our servers were up from 8am to 5pm.

Here's an actual true case: In a certain company the client paid to have their servers available 95% of the time. But one of the main servers was going down for a few minutes almost every single day. The support team was always able to bring it up within 5 minutes. However, the users were complaining a lot; many of them were actually losing their unsaved documents and the situation was causing a lot of frustration. Guess what happened at the end of the month? The server had a 96.7% reliability. It was always up at night and during weekends, when nobody was using it. What a terrible way of playing with numbers!

We discussed about creating specific procedures for different kind of systems in the first chapter. Likewise, the KPIs, SLAs and SLOs should be assigned to this system according to its specific needs instead of making a general statement. You must determine what is prime time in each specific case in order to adjust your metrics.

KPIS

Obviously, server reliability is not the only valid measure. There are probably dozens of different metric parameters to count in a company. Another tricky one is calculating KPIs for the service desk (also known as the help desk). One company decided to determine the performance of their SD department by the number of treated calls and closed calls per month. After calculating the first few months the numbers looked great! Most of the tickets were closed within a couple of hours, the majority of them within a few minutes. The number of calls to the SD had gone down, making people think the issues were getting resolved.

Unfortunately, nothing was further away from the truth. Even though the SD was congratulated by the management team, a little survey demonstrated a big fraud on the system. The SD personnel was misdirecting the calls and closing them. Sometimes they made a quick recommendation and closed the ticket. Some people were waiting on the phone for so long that they finally hanged up, and the SD closed the ticket. After quite some frustration with the service, the users stopped calling. They felt they had a better response by asking a colleague about the problem or calling a friend. So the final outcome was less calls and a lot of quickly closed tickets.

Consequently, we need to be very careful when creating the metrics for our service. As they say: "Be careful what you wish for, you might receive it". The real objective needs to be defined and additional metrics must sometimes be added to determine if the level of service has reached the desired objective. Just as we think of all possibilities when designing a security system, we must also verify any cracks in our service metric system. It should be fraud-proof. In addition, just like the IT management procedures, the metrics should be revised frequently to determine if they are still providing the desired information. We should never forget the main objective of our metrics: To understand the level of service being provided and appropriately provide the necessary corrective actions. It should never be used as a political tool or just as a sales pitch.

AUDIT FAILURES

Have you encountered a policeman or a traffic officer that will position himself just after the school zone starts so he can catch more people going over the speed limit in the school zone? We normally hate to see this happen, the main objective should be to avoid any accidents with the kids crossing the street, not to get more traffic tickets. In spite of this logical thinking, how is the agent's performance being evaluated? You guessed right, by the number of tickets. Unfortunately, his work is not properly recognized and doesn't match the main objective on why he is there. In consequence, he's taking measures to demonstrate that his doing good work, at least according to the numbers.

This same sad philosophy is also being applied in the IT field. We congratulate and give bonuses to auditors when they have big findings in their investigation. In consequence, their main objective is to find as many failures in the working procedures as they can. While this view-point will certainly stimulate them to do their work and look harder, it may also fail to accomplish the main objective: To have a good working system and make sure everybody is following it correctly.

With one of our clients, we assigned a certain person to do security check-ups on all the workplaces; we'll call him Mark. We wanted to make sure that everybody was properly keeping confidential information secured, computers were properly locked when users stepped out of their workplace and that nobody was writing their passwords on post-it notes (yes, there are sometimes still there, just below the keyboard). Mark did a first round check and came back with one finding, somebody left some important documents exposed over his desk. The manager properly gave a warning to that user and thanked Mark for his good job. After a couple of weeks, Mark came back to his manager and said: "Now you'll be very happy with me. I found five people that didn't conform to our security standards. Let's fire them!". He was all happy about finding more people missing the procedure. The manager was quite worried at this point. Of course, he didn't have the luxury of firing people, he needed those people. The real question became obvious: Why aren't these people following the procedure? Understandably Mark was not able to answer that, his assignment just consisted in finding the faulty people, not understanding the cause of the problem.

When performing audits, we should not miss the main objective. We want to make sure that people are following the right procedures when doing their work. The auditors should then be reasonable with their findings instead of looking for the missing crossed T's. I've seen auditors' findings where they judged the procedure was not being followed in a system change because the technician executing it didn't mark his approval! What a waste of valuable work! In this case the auditor was determined to make his quota, so he was looking for every little detail he could find. One factor for the auditors' evaluation should be the quality of his findings, and not just the quantity of them. But how do you determine how important is a certain finding?

Every finding should come with a reasonable explanation. In addition, the impact of each finding should also be clearly stated. This will give us a good idea of the quality of the finding. If somebody left bread crumbs on his desk, however inappropriate, it will not have a big impact on the organization; specially if nobody really has access to that person's workplace. In the other hand, if somebody leaves confidential information exposed on his desk, and other people who shouldn't have access to it may see it and use it, it may be classified as a high impact finding. Leaking confidential information may be disastrous for a company's image.

Just as important as finding problems during an audit is identifying their root cause. It's crucial to understand why a certain situation happened in the environment instead of just going to the quick solution and punish the naughty guilty person. In a certain place, changes were being executed without the proper approval. We found who was responsible of this action, and the person had knowingly executed several changes without following the approval procedure. The easy answer was probably to dismiss the faulty person because he was not following the procedure appropriately. However, when investigating deeply into the situation, we found out that there was a culture of putting enormous pressure on some people to have the IT changes executed on time. Upper management was worried about having some projects completed and delivered on time to the client. Replacing one person with another was not going to change the situation. If the same pressure was applied on the new person it will just be a question of time before he started to skip approvals and miss the procedure. In order to correct the situation, we needed to have serious talks with the upper management so they could better understand the importance of following the procedure when doing changes. Once the true cause has been determined it easier to provide the proper corrective actions that will really offer a valid solution for the problem identified in the audit.

All corrective actions should be applied with the original objective as the main guide. They should not be used to achieve personal reduction goals. They must especially help prevent any dangerous behaviours, identify obsolete procedures and assist people in better following the guidelines. An audit cannot be regarded as successful just because it has a lot of findings, but because the discovered issues found during the audit will help improve the procedures security and performance, re-assure stakeholders about the current practices and draw attention on missed opportunities. It also serves as a red light warning to help people avoid risky shortcuts.

INVENTORY

It is almost impossible to keep good IT metrics without a proper inventory. Most people will consider this quite obvious and you may even be asking yourself why is there a section about inventory? Don't all companies keep a good inventory of all their IT equipment? Sadly, the answer is no. Although most people believe they actually do. But when taking a closer look to what they call their inventory we normally find out that it's quite incomplete. People will often think that an inventory is just a list of equipment they own or control, but in reality it's a lot more complex than that.

Nobody will argue that you need a list of hardware equipment as a minimum. The first question I normally ask is how is that inventory updated? And the normal answer to that is that every time there's new equipment it gets added to the inventory. That's good, but how about when equipment is decommissioned? What happens when it moves from one place to another? Who's actually responsible for it and what is it used for? A list of equipment is absolutely necessary for financial purposes, but it's definitely not enough if we're talking about properly managing the equipment. We need to know where is the equipment at all times and how it's being used.

To start gathering information about your inventory you first need to consider what information you really need to keep. You may get some assistance by using some popular templates. However helpful, they were not created thinking in your own business. So you can base yourself on them and use some of their information, but you still need to customize it and create your own template, according to your specific needs.

Some people will assign names to the equipment for easy tracking. It's definitely easier to refer to an equipment as LondonServer1 rather than using serial numbers of some other weird denomination. However, naming equipment should follow a defined standard to avoid confusion. Some people will use the first 3 letters of the equipment's location, plus 3 letters to designate its function ending with 2 or 3 numbers; producing names like LONSER01 for example. Some other people will have some fun and use Star Trek, Star Wars or some other movie's spaceships, characters or places. Whatever your choice, it should give you enough selection to have enough names when your system grows and it must be used along all your IT environment.

Once you have your equipment's denomination standards, you can associate its characteristics with it. Equipment's model, make, serial number and any other permanent hardware features that may be useful to keep in the inventory. You then start adding some non-constant information, these are features that may change with time like physical memory, CPU capacity, network connectivity, etc. It may also be necessary to record the equipment's location and how to access it (console name, gateway information, etc.). This will give you a very good and important table of information about all your IT equipment, but the job is not finish just yet.

We often think about the inventory as the collection of information on physical goods, but that is just part of it. How about your soft goodies? Your software, programs, databases and applications are also important assets that need to be tracked. They are just as critical as your hardware equipment for the good functioning of your IT environment, but with one little caveat. Hardware equipment doesn't change that much, but your software inventory is constantly changing. You need to keep track of the software version you're currently using and keep that information always updated. In addition, software assets need to be associated with hardware equipment. We must document where is the software installed.

Here's an example to show why is this important. We received notification from a vendor that a specific version of their application needed to be patched for security reasons. It's not good to just know that you purchased a certain quantity of licenses for that software. At that point it was critical to know where was it installed and what version were we currently using. By keeping a good inventory, we were able to verify our IT environment and provide the recommended patch very quickly.

Finally, we also need to mark who is responsible for the equipment and for the software. This is quite critical when maintaining the equipment or when changes need to be performed on the system's software. The owner is not necessarily a technical guy, on the contrary, he will be a business person in most cases. We're not necessarily looking for who will repair the server, but rather for who needs it and will make sure the server is at the correct version and performing as expected.

We can never say that the inventory work is completed and over. It constantly needs to be revised and we should have the proper procedures in place to make sure that the inventory is always up to date. Servers may move from one location to another, people responsible for the equipment may move to other jobs. So without proper updates our inventory will just be a nice piece of history without much use. I'll make a confession here: a few years ago, I got a call from a technician that a hard disk was down on a server and needed to be replaced. The server was properly recorded in the inventory, but the rest of the information was totally out of date. We did phone calls, e-mails and broadcasted the situation to whoever could hear it; we could never find who was responsible for it. I'm pretty sure the server is still down sitting somewhere in the client's data centre. So you need to set-up a good procedure in order to always keep your inventory up to date, with useful and accurate information.

CHAPTER 4

Data Safety systems

Redundancy, backups and DRPs are classified as safety systems. You cannot ignore them if you're serious about your IT systems. Some of them may even be required by law in some countries were some kind of business continuity assurance must be provided to protect the clients. Even without these laws it's just common sense to have a good protection for your business.

A good risk analysis is necessary to determine the best mitigation procedures. Most of these kinds of analysis are based on two key factors to be considered for every possible event: the probability it will occur and consequences associated to it. However, when analysing the probability of an event, there is one additional factor that must be considered: Probability against the elapsed time.

HARD DISK REDUNDANCY

Let's start talking about hard disks to demonstrate why this is important. Hard disks nowadays are reliable. They are manufactory tested before they are shipped; so the probability of installing a faulty disk is quite low. Nevertheless, the probability of failure will exponentially increase with time. The failure probability may be quite low during the first few months, but it will radically increase after the first year and may even

become an almost certain event in 5 to 10 years. Therefore, when calculating the probability on these kind of equipment is more a question
or "when" rather than "if".

With today's cost of hard disk and good redundancy technology,
there is really no reason not to have a redundant system. Hard disks
will fail, if is not today they will fail tomorrow. Plus, there may be an
incredible amount of work to recover from a failed non-redundant
disk. Even if you have reliable backups, there is no guaranty that you'll
recover 100% of your data. Most people will install a RAID system for
hard disk redundancy.

We will not go deep into the technical details in this matter. But
we'll give a general description on RAID systems. RAID 1 is what
we call a mirror system. You must double the amount of disk you want
to make redundant. This kind of system is used mostly for Operating
System disk, or any other system that will require only one disk. So in
this case if you have a 500GB disk you must purchase another 500GB
disk to have it as a mirror of the first one. Any data written to the first
disk is also automatically written to the second disk. If one disk fails,
the other one will continue to work without interruption. In addition,
the failed disk can be replaced without shutting the server down. The
new installed disk will be re-mirrored with minimum effect on the
server's operation.

RAID 5 is commonly used when your data is distributed among
different disks. Users will only see one drive, like the C or the D drive.
However, the data on the drive is distributed among different disks.
For example, you may have a D drive of 2 Terabytes (2000 Gigabytes)
composed of 4 hard disks of 500 Gigabytes each. Mirroring this kind
of configuration may be costly, so that's why RAID 5 is created. This
is a system where your data is written in a redundant way on all your
disks. You still lose a percentage of your disk capacity, but it's still better
than 50%. You may have 3, 5 or more disks working together as one. If
one disk fails, the other ones will continue to work without a problem
and will avoid any server outage. Just like a RAID 1, the faulty disk

may be replaced without shutting down the server and without any service interruption.

Nevertheless, you may run into serious trouble on either system if 2 disks fail at the same time. Therefore, faulty disks must be replaced as soon as possible to avoid any serious danger. Change management procedures must be written according to this real possibility. The subject is covered on the Change Management chapter.

SYSTEM REDUNDANCY

Unfortunately, disks are not the only thing that may fail in an IT system. The servers themselves may also fail and the causes may be quite diverse. Faulty memory, CPU systems and other server components may cause a hardware failure that will affect your server. However, applications and badly configured Operating Systems may also cause a server to hang and stop working correctly. Consequently, we also need to evaluate risks and plan for redundancy when building IT systems.

Similar to hard disks, you may also plan to have servers working together so if one of them fails the other ones can pick the load. Server failures are completely unnoticed by the end users. In order to achieve this goal, we can have servers working in clusters.

There are several kinds of clusters to support failures. An active-passive cluster will have a working server getting 100% of the load while the other one is standing by. They often will share the same data store. In this case when the active server fails the other one will recognize that there is a problem and will automatically active itself and connect to the data store. When the operation is successful it may be completely eventless for the users. In some other cases the servers will have their own storage units continuously replicating with one another to keep the information mirrored on both servers.

This configuration has the advantage of being simple and easy to manage. You know exactly where you're going and where are the services on processes running at all times. However, it can be quite costly to have a second server just sitting there doing nothing; waiting for the next failure from its brother. You also have to make sure that any configuration changes are also done on the clustered server to keep them working at similar capacity. Moreover, data storage access (or replication in case of duplicate data storage) should be continuously verified.

Some IT architects may opt for an active-active server cluster. In this case both servers are connected and sharing the load from the users. This configuration certainly increases the server capacity as you can double your infrastructure to better respond to customer demands. The biggest problem with this solution is that server capacity may frequently be over 50% on each server. Alarms are not normally triggered since it's considered as ordinary. But when a server failure does occur, the clustered server may not have enough capacity to handle the full load and you'll experience an outage at that point; which will cancel the usefulness of the cluster as a safety feature.

One of the most common cluster configurations is to have several servers, and not only two, working as part of the cluster. It can be built in an active-active-active-active configuration or in an active-active-active-passive configuration (I'm using four servers on this example, but there can be any number). This is often called a farm of servers. In the first example they all share the load and they may all take part of the load if one of the cluster members fails. However, you must still make sure they are not working near to their limit capacity to avoid any outages. In the second configuration, there is an additional server standing by just in case any of its 3 brothers fail. The ratio in this case is better than having a one to one cluster failover where you have to double your server capacity.

VIRTUALIZATION
Going further on this solution is server virtualization, where servers are not directly attached to a physical machine. As a matter of fact, one physical machine may host several virtual servers. Users will access

these servers just as if they were installed on different physical boxes; in other words, they will not see the difference. This kind of IT architecture offers several advantages, especially when you cluster different hosts together. When doing that you may then transfer virtual servers from one host in the cluster to another. This characteristic lets us better balance the load among the different hosts in the cluster.

Other than the redundancy advantages in using virtualisation, there are other great benefits. When using a physical server, you need to make hardware changes each time you need to add more memory, CPU capacity and sometime even storage capacity; unless the server is attached to an external storage unit like a SAN. Once you're using virtual servers, these elements are easily configurable and can be changed by simply modifying the server's virtual configuration. Some of these modifications can even be done on the fly, while the server is running and with no interruptions. This can save a considerable amount of time when adjusting server capacities to its requirements.

We cannot talk about virtualisation without having at least a short discussion on Cloud computing as virtualization actually provides the basis for it. Let's just say that for example that you don't want to manage your storage and you want to give this responsibility to somebody else. You may purchase Cloud services from a provider and voila, you now have your magic storage. You will not care how many servers are being used to store the information, or on which operating systems is your data being stored. It will not be your responsibility anymore. You just have to pay for the amount of data you're using. You can do the same with CPU capacity, virtual servers or platforms or any other IT service. There's a great variety of services being offered by different providers. In addition, you may also have an internal Cloud that you manage yourself within your own IT environment using virtualization and automation.

We will not discuss this subject in detail on this book. For now, will just provide this short explanation. You may want to do some additional reading if you feel this is something that may help you IT environment.

BACKUPS

As much as you want to safeguard your system, add redundancy and make sure there are copies of your data everywhere, you will always need to have a good backup system that you can trust. There are many different types of backup systems out there, each one with its own virtues and handicaps. You must choose the one that better adapts to your organisation needs and requirements. Every IT system is different, so you need to consider certain factors when deciding which one to use.

Some systems can backup your whole environment in a flash, very very quickly. But how fast can you restore from them? And when you do need to restore, will you be restoring the whole server or specific files? Who are you putting in charge of restores? Will they be done by your IT professionals or will that be done by the users themselves? These are some questions that may help you determine which system better adapts to your needs. Picking the best in the market is not enough, it might not be the right one for you.

You will also need to document your backup procedures according to your needs. For example: For how long are you keeping your backups? Not only are there practical times and dates on this subject from a technical point of view, but you might also need to consider if you have any legal implications when keeping your data in backup. It also needs to be determined if there is a need to store the backups outside your facilities. You don't want to have a general rule for it, you may need to keep some information out for disaster recovery procedures; but that doesn't mean that all the information on the backup needs to follow the same rules.

Keep in mind that data may sometimes be very valuable, and there are many people out there willing to take some risks to obtain your data. It must always be protected, especially if it's leaving your data centre. The information must be encrypted, regardless if it's contained on tapes, hard disk or storage cards. It should be transported by security personnel and stored securely. Unless you are just sending out public

information, which is rarely the case, you must follow strict security rules.

In addition, there must be a record of where are the tapes (or other
backup storage devices) at all times. Every tape (or other storage device)
has to be accounted for in an inventory. There should be a strict procedure to determine who has access to the tapes and how they can travel
from one place to the other. Unfortunately, we hear from time to time
about companies losing a tape, a disk or even sometimes a laptop. You
don't want to make the news that way so a good procedure is essential
to make sure the information is protected.

MONITORING

Finally, we'll talk about monitoring as a good safety tool. If you're serious about maintaining your IT systems working well and being healthy
you must consider monitoring as a critical tool. A good monitoring
system will not only warn you when something goes wrong, but it will
also inform you whenever there are any out of the ordinary events. It
should also give you enough information to analyze if you IT environment health is deteriorating before any big outages occur.

Just as any other IT product out there, you must evaluate what is
available on the market and which product will better fit your needs.
Unfortunately, the price tag is often the big winner when deciding
about the system. It's sad to say that many people will come out as
heroes as they successfully completed their projects installing miserable
monitoring systems at low cost that really don't serve the purpose.

For example: I have worked with many companies on which their
monitoring systems will not warn them whenever a server goes down.
The stupid monitoring tool will only send the alert once the server
comes back online. We often have to answer the question: "How come
nobody was alerted that the server was down?". On which we normally
answer: "Because the monitoring tool is installed on the server and will
not report when the server is down". As absurd as it sounds, nobody

ever did anything to correct the situation and replace the monitoring system. Everybody just consider it as part of life's normal troubles and left it the way it was. Nobody dare to question the choice of monitoring tools being used.

The truth is that with today's technology this doesn't need to happen. There are many good monitoring systems out there that will actually warn you in real time when there are problems. These are normally systems based on a monitoring console that will call servers and gather information about their health instead of relying on the servers themselves initiating the communication. However, I must say, this may be more difficult than it looks at a first glance. The IT environment is not perfect and glitches do occur, and quite frequently. So after you choose the right monitoring tool for your environment you then have to take the time to set-up the correct thresholds to determine on which situations you want to get alerted.

One of the biggest mistakes is to say: "I want to get alerted when any problems are detected". Really? Do you really need to know if there was a 3 millisecond glitch on the network that didn't affect anybody? I agree that may be useful information to analyze when studying the performance of your network, but we're talking about alerts here. The important issue is to avoid crying wolf. When you have too many alerts, people tend to ignore them. In consequence, when you'll actually have a major event, people may not pay the proper attention to it and delay their response. The monitoring system needs to be fine-tuned so it actually alerts about urgent issues and not on common events.

Another common mistake is to apply general monitoring tools for the whole environment. For example: Thresholds may be set to send warnings when CPU utilization exceeds 95%, or when hard disk capacity is close to 90%. What may be a critical situation for one system may be absolutely normal on the other one. You may want to be careful with CPU utilization on systems that have client on-line access as this will normally translate to a lower response time for the users. On the other hand, you want to use your full CPU potential when running a batch process on a machine that doesn't do anything else. A good solution for this is to create different alarm profiles and then assign them to your different systems and applications according to your needs.

Once the thresholds and proper warning are configured, now you need to determine how to manage the alerts. You may end up with a big list of alerts, going from hardware issues, OS errors and application problems. It's obvious that nobody wants to get called in the middle of the night for simple issues and everybody will push for someone else to take the first look at the incident. The rule of thumb to follow is: Whoever is in a position to solve the problem should be the first one contacted. Some organizations will like to avoid the controversy by sending the alerts to the help desk and have an intelligent person analyze the alert and contact the right people for it. This method is widely accepted. However, you do have to sacrifice some response time with this technique. If you're willing to pay the price, then this procedure may be acceptable for you.

On the other hand, sending all alerts to a single support group is normally just a waste of time. It ends of to be quite frustrating for the selected support group to sort out all incidents reported by the monitoring system. In addition, there is always a waste of time when the resolver group is not contacted first. Every support team has to be responsible for their own work and will normally be the first one contacted to repair or improve the configuration under their control.

A common error is to group monitoring incidents by type and then assign them to the support teams. It should be the resolver, not the associated group that should take responsibility for the incident. For example: You may be tempted to setup alerts for hardware issues like available disk space, CPU capacity, etc. and assign them to the hardware/OS support team. But if you're running an application on the server, and the hard disk used by the application team is the one filling up, there is not much that the hardware/OS support team can do about it. The resolver team in this example will be the application team.

As much as a good analysis is done, there will always going to be alerts going to the wrong team. A periodic analysis should be done on the monitoring system to determine if the alerts have been correctly configured and to analyse if you have a good percentage in hitting the resolver team when assigning the incidents. As your IT system evolves, so should your monitoring system. What was right a year ago may not be the correct answer now. The monitoring system should continuously be configured and adapted to the most recent requirement of your IT environment.

CHAPTER 5

Incident Management

"The system just failed, we needed back NOW!". Nobody wants to hear these words. Sadly, this is one of the most common phrases in IT management. Of course, the best incident is the one we can avoid. That's the reason why proper redundancy and monitoring as so important. However, let's face it, incidents will always be part of our IT life, just as snow storms in Canada or rain in London. We just need to create a good incident management procedure in order to restore service quickly and efficiently (and may I also add painlessly).

What is the first priority whenever we experience an incident? Most people will answer that the number one priority is to restore service. People who depend on the failed system will strongly recommend that. But how good is it to restart service if it will just go down again in a couple of hours or even in a couple of days? Knowing what is going wrong and why we're experiencing an incident is also quite important. Should it be our first priority? Well, that wouldn't make sense either, will it? You cannot spend days studying the situation while your system is down.

CREATING A GOOD BAL-
ANCED PROCEDURE

Reasonably, a good balance is necessary between putting the efforts to restore the system and understanding what is the problem. One of the reasons why some people hesitate to restore a system right away is that some of the evidence might disappear once the system is up and running again. Nevertheless, this is clearly not the time to analyze every piece of information that will help us prevent another incident. The focus should shift to the effort to restore the system asap once we gathered the necessary information.

But what is the necessary information? The real and honest answer is that we may not know that during the incident. You may also not be able to think about every little detail at that time (believe me, your internal or external clients will not let you do that). All of these questions have to be answered before the incident occurs, while writing the incident management procedure rather than trying to improvise what to do during the crisis.

In consequence, a good incident management procedure should not only have the steps on how to restore the system, but also on how to obtain the important information that will help us understand what happened in order to prevent the problem from happening again. On the other hand, the procedure should not be so complicated and cumbersome that it will take too much time and delay the system restore. A good balance is critical to effectively run incident management. How can this be accomplished?

CUSTOMIZE PROCEDURES

As mentioned in the first chapter, we cannot just create a general incident management procedure that will cover all our IT systems. You may want to follow ITIL or some other good established procedure in your company, but be careful. Even though general guidelines are very important, every single system has particular features that must be treated in a unique way. If you use the airline industry comparison again, it will be like running a general procedure for all airplanes and asking Cessna pilots to make sure the landing gear is down before landing. Remember, once people identify a procedure as being an obstacle rather than a tool they may start ignoring it and cutting corners to make their work more efficient. This may look like a good idea for some people during some time, but at the end it can become quite dangerous.

One of the main reasons why you may want to customize your incident management procedure per IT system is application knowledge. You may have a good IT team of people working on your systems who know exactly how things work and how to recover from any situation. Unfortunately, they will not be there forever, and once they're gone, their knowledge will leave with them. Unless, of course, we manage to write down their knowledge and have good documentation.

A customized procedure should include information on what is the right configuration for this system. Having 10% of free disk space may be normal for some systems but not for some others. So, when a system is not responding, what should we look for? Memory, disk space, CPU utilization? A good tailored system will help us gather the right amount of information without wasting time in some other factors that are not affecting this particular system.

Some other aspects to be included are the particular steps to recover the system. Do we need to shut down a database before restarting the server? How about any applications or special connections? Not following the right steps can cause even more problems and considerably delay the recovery time. For example: it may take hours to restore a corrupt database because it was not shutdown properly. Taking just a couple of extra minutes may save a lot of unwanted work. Some people may argue that a good technician must know about it and look at these things. Maybe, but not all technicians will think the same way. If all steps all well written, you reduce the possibility of somebody forgetting an important step.

Another important element to be considered is connectivity to other systems. How may this application failure be affecting others? Should we be contacting other people or checking other systems? In today's IT world, quite rarely will a system be totally isolated. It is a common mistake to recover a system without fully verifying the consequences of a system breakdown. It's kind of finishing to make a cake and leaving a big mess in the kitchen (people with cooking teenagers will totally understand this example). A good system cleanup is crucial to make sure all systems are OK. Sometimes other teams will only find out the next day that there was a problem that affected their system. Any other connected application should be verified as part of the recovery procedure to make sure everything is OK before dismissing the incident management team.

INCIDENT COMMUNICATION

"The big crisis moment is here now! Red alert! Everybody on the bridge!" So we have Joe, George, Tony, Miguel, Anna, Tracy, Peter and Rebeca. Good! Now, how many people are really working on the incident? Well, usually you have one, maybe two. The other ones are just following up or waiting for their turn to intervene. This follows the general idea that whenever you had a crisis you need to call everybody and have all hands on deck. This practice is widely used to reassure everybody that we have everybody working on the problem. But in reality, is this the best way to solve an issue?

First of all, we need to define who must be contacted on a need to know and need to act basis. Do we need an incident manager? In most cases it is quite useful to have somebody organize the chaos. Unless you have quite a small IT team and everybody can work by themselves and follow the procedure, you need someone who will supervise the whole operation. He doesn't have to know all the systems, but he will be able to follow the procedure and make sure everybody else does the same. He will also be responsible for communicating all updates about the incident to the right people. This means that not everybody that needs to be contacted must be on the bridge. We'll have three contact categories:

First, we need to contact the action takers. These are the system administrators, databases administrators, hardware technicians, application experts, etc. We find the problem solvers in this category, the people that can get their hands dirty, do some clicks on their computer and fix the issue. A bridge without these people is only a meeting, we can talk a lot about the incident but nothing will get solved unless they're present.

How about the IT service managers? It's a normal ritual to call them at the same time as the technicians, sometimes even before. The big question about this is why? Really, why do we need to call them? They are normally on the bridge to make sure the incident is on its way to get resolved, procedures are being followed and to get the technicians to follow-up and do their work correctly and responsively. Isn't that the job of the incident manager? This is like having a second captain on the bridge. They are often called because a) we don't trust the incident manager, b) we don't trust the technicians or c) we just don't trust anybody on the bridge and want to make sure things are well done.

Regrettably, some people believe that incident managers are only there to open the bridge and call the right people. They think they need a service manager to get the work done. This is just adding a second supervisor because we don't trust the first one. Work should always be well done by the people responsible for it. This applies to the incident manager and to the technicians. They should be responsible knowledgeable people capable of doing their job. Other people may be called only if needed and on special situations, but not as a routine. Having too many people asking questions on the bridge will not solve the incident any sooner.

In second place, we find the people that need to know now, but don't necessarily need to be on the bridge. This may be IT directors or managers that are responsible for the IT environment. They need to know that something is not right, but they have nothing to do on the bridge. However, these people cannot wait until the next morning to find out about an issue, they must be directly contacted. Even though their presence is not required on the bridge, they may eventually login to follow-up or directly get a status on the incident.

In the third category, we find people that need to know there is, or there was a problem. These may be end users, stakeholders or people that may be related to the failed application or service. They don't really need to be contacted directly and we definitely don't need them on the bridge. However, it's important to keep them informed about the situation. Not just that there is a problem, but also include additional details; as much as possible. They need useful information like what systems are being affected, what is being done to solve the issue and when can we expect service to be restored.

CONTACTING THE NECESSARY PEOPLE

The contact method will then depend on the category of the contact. Let's face it, people are not normally waiting around waiting to get called. They might be busy with personal activities, sometimes out of the house or even sleeping. This 24-hour availability idea works if we understand we're calling actual living human beings, not machines. So how do we alert the necessary people?

Pagers are widely used in some environments to contact people in the first and second category. It used to be a good communication method, but technology has evolved and our communication strategy should also evolve with it. I just can't understand why people will use a numeric pager or send a numeric message. This is an obsolete one-way communication that doesn't say anything, just that there's a problem and that somebody needs to call back. We don't know when the person reads the message or if he's available. We just need to sit down and wait for them to call us back so we can explain the situation. This can be a long wait when we have a critical problem in our hands.

How about text pagers or text messages to mobile phones? They're better than a numeric message because we're at least sending some information like: "Montreal1 server is down, not responding. Please call the bridge at 1-800-etc.". This way the responsible technician will know what is the problem and can even start preparing while login to the bridge. Another advantage of this method is the possibility of doing a quick broadcast to all parties involved. Nevertheless, we must consider some issues with this method. If we're using pagers we should be using at least a two-way pager so that people can acknowledge that the message was received. Otherwise we'll just wait for the contacted person to call the bridge. On the other hand, if we're using mobile phones we must find a way for the person to be clearly alerted by it. With dozens or maybe hundreds of text messages being received per day he may not think our message is critical and it may take some time for him to confirm it.

One of the best ways of contacting people are direct calls to their mobile phones. The advantage of this is an instant response from the contacted person. We get to talk to him right away and explain the problem. In addition, the contacted person can quickly give a feedback like making recommendations, suggesting to have an additional person on the bridge or even minimizing the alert if necessary. However, this works well when you just have one or two persons to contact. It's time consuming when you have to directly talk to a big team of incident resolvers. It may take approximately 5 minutes per person, which can be a lot if you have a big team of people to contact. Besides, sometimes you may be talking to a voice mail recorder.

So, what is the best way to contact incident resolvers? This may depend on where you leave and the culture of how people like to get contacted in your technical team. You may want to use parallel methods like sending a text message and confirm with a direct phone call if a confirmation is not received quickly. In other words, do your best to have everybody working on your incident as soon as possible: phone, text, pigeons, smoke signals… whatever works best for you. Now, should we just bring everybody to the bridge?

INCIDENT START-UP

Now, how do we get people to start working on an issue? It may all depend on how the issue is actually identified. We'll talk about three incident starters: an IT technician, a monitoring device and the end user.

IT technicians will normally be accessing the system. They will login to servers and verify the systems from time to time. It will often happen that they will identify a high severity problem as they might be the first ones to see it. It should be clear to them that whoever or whatever identifies an issue must start the incident management procedure. If he has the right access and knowledge to solve the issue he must proceed with the proper actions. However, if it's a high severity incident then he must contact the incident manager, even if the issue is already solved or will be shortly.

When a monitoring device finds a problem, it should automatically contact the right support team to fix it as explained previously on this book. A common mistake is to have the monitoring device contact this one team who will look into the issue and then call the appropriate team to solve it. Most monitoring systems nowadays are intelligent enough to contact different teams for different kind of problems. Although they may not be perfect, it's better to have a good first shot at the right team rather than missing the opportunity to solve the issue quickly.

Unfortunately, it is sometimes the end user who finds out there is a problem with a certain system. Normally he's not in a position to understand if there is a server problem, an application problem or a network issue. However, he needs to have somebody to contact. Normally we create a service desk (or help desk) to receive calls from end users and route them appropriately. The service desk personnel must have a script that they can follow in order to ask the appropriate questions to the end users and better direct the call for a proper resolution. The answer may be to call a certain support team or even to call the incident manager. A good response script with the right questions can certainly be a useful guide for the process.

AVOID THE CROWDED BRIDGE

"Everybody out of the room, the doctor is about to operate." Remember that phrase? It's typical to hear the doctor keep only the necessary people in the operation room, and the reasons are quite obvious. You don't want to have the family members in the room to avoid an emotional environment; that can make the doctors lose focus on their objective while operating. You almost never have the hospital director in the room either. He will normally not attend the operations unless he has a serious concern about his staff.

We must keep a similar principle on the incident bridge. We will keep the people that we really need there in order to stay focus and solve the issue quickly and efficiently. That means that we should not have all stakeholders on the bridge. If for any reason the IT managers, CEOs, clients or anybody else needs to get into a meeting they should open a separate bridge for their discussions and let the technical team complete their work. Just as somebody will come out of the operation room from time to time to provide an update, the incident manager must make sure that appropriate updates are provided to all the stakeholders.

On the first chapter we touched the notion of the crowded bridge. We kind of have the feeling that we need all IT experts there to make sure the issue is getting resolved. The normal practice is to have the OS support expert (normally a Windows or Unix system administrator), the hardware expert (in case there is a hardware issue) and the database DBC (to check on any database problem if one is being used), the application expert (whoever can give us some information about why the application is not running and what is needed).

In addition to this we normally add all the managers and supervisors: the incident manager should normally open the bridge, make the introductions and follow-up on any actions being taken. He should also log and send the necessary documentation. Besides him or her, normally we have some kind of IT manager, making sure the technicians are working at par with the problem. Quite frequently we'll have the manager responsible for applications running correctly. On top of all of that, it's common to have some other managers logged to the bridge to catch up on the latest news and follow-up on the incident.

I'll deviate from the normal practice of having all of these "experts" on the bridge. I know that some people may not like this, but let's just go with two basic principles. The first one is just common sense, let's just have the people we need on the bridge to solve the incident. Having more people may delay the incident resolution. My second point is: let's give people enough access to solve the issue. I'll explain…

"The application is down; everybody is on the bridge!" Well, almost everybody. The application team has quickly identified a problem and called the incident manager to open up the bridge. It's a known issue dealing with some memory problems and the application team quickly recognized it. However, they cannot proceed with the solution because it involves a server reboot. So a bridge is called and the incident manager is informed that a system administrator is required. He pages one from the list… 15 minutes later there is no response. Well, we'll page him again! 10 minutes later, still no response. Now we page the system manager, he'll get us someone. 5 minutes later he comes to the bridge and the incident manager explains him the situation. He then drops from the bridge to try to get someone to help. 15 minutes later we finally have a system administrator on the bridge and the incident manager explains the situation once more. The system administrator now asks for some time while he logs in to the server and starts the reboot. About an hour and a half later the server is finally rebooted and the application team confirms the service has been restored.

The logical question should be: "why didn't the application team just reboot the server?" Yes, it sounds logical, doesn't it? Nevertheless, most people in today's IT world will quickly say "NO, that's a big no no! Application programmers should not reboot the server, that's the system administrator's job." Really? Why? We have a problem in our hands and we have to wait for the "system expert" to reboot the server? Even my kid knows how to do that!

The big argument is that, once we engage an internal or external team to support the servers then nobody else is allowed to touch them. It's normally said that they cannot support the machines if other people manage them. They also say that if other people have manager access to the servers then they cannot be responsible for them. We discussed this subject on the second chapter when talking about specialized technicians, but I'll add a few m

These arguments are not strong enough to support this idea. Moreover, they should not be used as an excuse to delay proper and quick incident resolution. Additionally, the arguments are not completely true. First of all, there are plenty of activities going on in the server. Processes are being started and stopped, disk space and memory utilization varies a lot and even some external factors like network connectivity are beyond the system administrator's control. To top it off, reboots may even be common on a server. Some systems will reboot automatically after a server crash, a hardware failure or even after a system patch. So it's nonsense to allow this to happen automatically but not allowing good technical professionals to reboot a server manually.

Secondly, if we have good access practice then we know who is logging on the server and doing what. We should be able to know if any changes have been made to the system and who did them. System administrators actually go to the log file to trace any changes on the system, so they are able to know if anybody changed something important on the server.

In conclusion, we don't need to wait for an expert to reboot a server or do any other basic tasks on the system. We should give some teams, like the application programmers, enough access to troubleshoot and correct the problem without having the Brady bunch on the bridge every time there is a problem. Some of these tasks may include: rebooting the server, start and stop some services or processes, start and stop the database, check system settings etc.

What if the application team in this example doesn't want the added power and responsibility? I remember this application team which were masters of diversion. Every time their application was in trouble they asked for the old cranky server to be rebooted. They always called the server support team and told them that the server absolutely needed a reboot. The old server always had a lot of trouble rebooting, it could take hours sometimes. After the incident, everybody was so mad at the server support team for taking so long that they totally forgot that the real problem was the application, not the server.

People need to be responsible for their systems and stop the blame game. The application team will often blame the server because the CPU or the memory reached 100% utilization. Well, isn't it the application using these resources? On the other hand, the server support team has to be pro-active, check their logs and server history and warn the application team if they see a trouble trend in resource utilization.

But what if the application team reboots the server and it doesn't come back? Well, now you may have a good reason to call the server support team. You may need somebody with more skills to check if there are any problems on the server. However, you may also solve the issue by giving even more power to the application team or having somebody press on the power button.

After the server reboots, the system administrators can still check the server logs and troubleshoot any issues on the server if any. But again, in this case, if the application team has already identified the issue and know how to solve it we may even avoid this step.

You may have some trouble selling this idea at the beginning. Most people will like to work in the traditional way not to get in trouble. Nevertheless, giving more access and power to other technical teams was the original model, before the specialized technicians. We used to work that way before. Specialized technicians still have their place in the IT world. It is still very important to have specialized people identify issues on servers, databases and other resources. But they should be used for exactly that, and not for trivial tasks that any general technician can do.

This underlines the importance of having a good incident management procedure. You may specify who has access to do what and which limits are you imposing to each team. You may also define certain recovery procedures that most technicians can follow without being an expert in other fields.

DEFINE THE SYSTEM'S CRITICAL TIME FRAME

Here we go, another alert! Everybody is on the bridge again. We have about 10 people, all of us wasting a good part of our weekend to solve another incident. Almost everybody except one, the hardware technician. So here we are, talking about the weather and trivial stuff until the incident manager comes back to the bridge will an ironic message: "The hardware on these server is not supported during weekends. We need to wait until Monday morning to call for support!" What a waste of time!

The incident management procedure should comprehend the notion of critical time for each system. Some applications need to be working 24 hours a day, but not all of them. Support procedure and contracts should be done accordingly. There is no sense in saving money on a contract by having just weekday service on a critical system. On the other hand, it doesn't make sense to have everybody working on a problem off hours if the system is not critical.

When defining support contracts for our systems we should always check this aspect and consider all possibilities. We normally check application and server support, but how about other external teams? What if a hardware replacement is mandatory? What if the support teams need to call the vendor to troubleshoot an issue? Trying to cut costs on this field is like buying car insurance for weekdays only. What if we ever decide to take our car on the weekend? Contract should be done following our needs and we should be ready to justify and defend the service we need in order to provide a professional response.

KNOWLEDGE BASE DOCUMENTATION

Have you ever seen a house that was on fire after the fireman crew leaves the place? It's usually not pretty. Even though the fire is over and everybody is safe now there is still a lot of work to do. The emergency team's task is very important, but there is still a lot of work to be done once the incident is over.

The first piece is probably documentation. We absolutely need to document what happen, what was the incident about and what was done to solve it. All interventions, actions and even comments from the support teams should be clearly documented and kept in a knowledge database. All this information will not only be useful to resolve future incident but it's also critical for the RCA (Root Cause Analysis).

As much as gathering all the information in a database is important, having the right access to it is equally critical. This is a popular problem in many databases. Even though they have all the information about what happened, it's stored in a way that is very difficult to access it. Most of the time you may access the event by incident number of by date. Frankly, who remember an incident number after a few months? Or the exact date an incident happened? Losing access to the information is just as losing the information completely.

Information in the incident database should not only be accessible by incident number, but more importantly by any other means an incident resolver may need in order to relate the incident he's working with the history. This should include searches by server name, application, system and special resources.

As usual, access must be granted on a need to know basis. At least, anybody that is working on a certain system should be able to search any history information on the systems he manages. Other people like IT managers should be able to consult the database. Data should be immediately accessible to anybody that needs it. When people are troubleshooting an incident, there is no time to search for tapes or look at reports.

ROOT CAUSE ANALYSIS

It is critical to understand what caused a system failure or malfunction. As important as it may be to recover the system on time, it's equally important to have a good understanding of the situation to avert any future crisis situations. The incident manager is responsible for documenting any actions taken during the incident as well as who performed each action. All the people involved should then analyze the situation and present it during a follow-up meeting.

The root cause should go deep into the issue. The answer should never be something like: "the server rebooted", or "there was a memory overflow". Even worse, by no means should it ever state something like: "this kind of things happen". We need to understand why and keep asking questions until we really get to the real underground cause of the incident.

For example: Was the server rebooted? Why? Because of a memory issue. So next we ask: "What caused the memory issue?". For that we may get an answer like: "The application went into a loop and used all available memory". So then we have to keep on asking why until we finally understand what triggered the incident.

Just as important as knowing what happened is taking the appropriate actions to mitigate our findings. Otherwise our RCA will just be good historical information, but will not completely solve the issue and we're bound to live the same experience again. For example: let's say we had a hardware issue, like a hard disk failure. There is not much we can do to prevent it, but we can mitigate it by adding redundancy so the next time it fails it doesn't affect our system.

Working on an RCA is not the time to play politics. We should not tailor our recommendations and actions to accommodate the client's budget, ideas or IT philosophy. Those things will change with time; however, the facts will stay the same. For example: If the client or end user doesn't want to pay for redundancy that's his choice. However, the fact remains that adding a second disk will avert the crisis. Maybe in the future the budget will change, or the person responsible for the system will no longer be there. Stating the facts in the RCA is crucial and must remain honest at all times.

It's often said that "an objective without a date is just a dream". When documenting the appropriate actions to mitigate our findings in the RCA we must assign them to a responsible person and add a target date to complete the action. While running the RCA meeting, it's useful to have a grid will all the activities, who is responsible for each one of them and their expected completion date. Now, if the assigned person is not in the meeting he should be invited on it to make sure he understands why the action is being taken and all necessary discussions can be concluded. Once the grid is ready we should have somebody assigned to follow-up on it and make sure all actions are properly completed by their target date.

CHAPTER 6

Problem Management

So, what is the difference between problem management and incident management? Basically, an incident is an event preventing your system from working correctly. A problem on the other hand is an issue that may cause multiple incidents, can prevent people from working on a system at its full potential or than can eventually degrade your infrastructure.

ITIL clearly defines a difference between problem management and incident management. However, most people will focus a lot of attention in solving the incident quickly while neglecting good problem management. Many people are mistaken by the idea that by doing RCA follow-up they are already doing problem management. Although RCA actions may be part of it, good problem management it's quite more complex.

PRO-ACTIVE PROBLEM MANAGEMENT

Most of the problems are created following a user complain. People may be unsatisfied by a system response time, or could be fed-up that a certain system is frequently unavailable. That will normally get people to start an investigation and find appropriate solutions to address client demands. The problem is then documented with the actions needed to correct it. All information about IT problems should be kept in the problem management database and must be accessible for consulting.

Nevertheless, we should not just wait for problems to get reported to start the process. A good pro-active approach is necessary. It helps prevent incidents and keep a good running environment. It's also useful to identify future IT architecture environment needs and have a good knowledge of current resource utilization.

With that purpose in mind, incident reports should be frequently analyzed to identify any problems in our system. Is there a specific server having trouble? Is there any application that is causing grief? Is incident resolution time adequate? The answer to these questions may trigger new problems to be studied.

Other sources of information may be provided by the monitoring system. How are we doing with resource utilization? Do we see any disturbing trends? The information should be continuously analyzed to determine if our IT system is healthy or if any intervention is required. This is similar to your annual medical check-up (although it should be more frequent on IT systems). The doctor will run some tests and check the results so any serious problems can be caught and resolved early. Likewise, monitoring systems should provide us the required information to evaluate our IT system's health and identify any important issues to be resolved.

In addition to all these sources of information, you should also take some time to review the RCA database. Are all actions being worked at as agreed? Is there any dangerous trend when we analyze the root cause of the latest incidents? And quite important, are any recommendations being dropped? Unfortunately, some good ideas may be lost due to lack of interest or limited resources. A good analysis in RCA history will teach us if any bad behaviours have been adopted by our staff and apply the proper corrective actions to fix it.

RISK ANALYSIS AND ACCEPTANCE

So we found the problem, and we may even know how to correct it. Now comes the fun part, trying to get the funding in order to start the project and fix it. Unfortunately, not everybody agrees when and how resources have to be spent to correct a problem. Therefore, besides uncovering an issue that needs to be resolved, we must also have a good analysis of the situation so we can prove our argument.

A deep and meaningful risk analysis should be done to explain and document the situation. In other words, what will happen if we don't do nothing? And even with a time lapse, what will happen today? Tomorrow? In a year from now? This will provide a good idea of the urgency of the situation so the finance department can better plan for the budget. Which leads to the next point: whenever possible, include the financial problems that can be encountered if no solution is done to correct the problem. Some people may not necessarily care if some workers are un-happy and not comfortable with a certain issue. But it may be different when we show the economic impact of the situation.

As much as we can explain how bad is the problem or how important it is to solve it, we may not be able to convince all the stakeholders about the importance to take action and fix the situation. In that case the problem should be correctly documented, with the risks that go with it, and it should be accepted by the responsible person. This will not only protect you if the problem grows bigger; I'll say it bluntly, it will make the responsible people, responsible. It's easy just to deny the problem, or try to make excuses. It may be a different story when you have to sign a document to confirm that you're accepting to take the risk for a certain situation.

It's obvious that people will avoid signing risk documents, so you have to establish a procedure of awareness and negative acceptance. For example: If I can prove that the situation was published, and that the right people were contacted about it, this may represent their acceptance; even if they haven't actually signed anything. It still means that they were made aware of the issue and that they decided not to act on it, it's almost just as good.

BENEFITS

One of the obvious benefits of running a successful problem management is that we get to know our IT environment much better. We're more aware of any trends and issues. There is a considerable reduction in the number of incidents and emergency changes as a result. Any amount of time spent on this exercise is a good investment.

Good problem management procedures are also great as business cases to start new projects, modernize the IT architecture and improve the infrastructure. It's easier to demonstrate the need to buy a new set of servers if we can show the old ones are always giving trouble. It's also great to demonstrate how a new architecture can solve some existing problems and improve IT services.

CHAPTER 7

Change Management

How will you like to use your good old computer from 10 years ago again? Forget it! There is no way! Technology evolves quite fast and changes are the norm. Not only hardware has to be replaced, but the applications also need to be modified to use the newest and finest technology.

It is clear that changes should always be performed in an IT environment. However, without proper control we may risk to severely affect our business. It is crucial to have good change management procedure that will help us better manage all the changes going on in our environment and make sure they're following a good discipline.

Many companies have decided to adopt ITIL or some other kind of universal guidelines. This is an excellent strategy and certainly helps keeping things under control. However, just as with incident management, change management must also be tailored to your own needs.

Change management should not be used as a change stopper. Remember, we must not stop changes from happening. Nevertheless, we want to avoid conflicts and we also need to document any changes in our environment in the change management database. This information should be easily accessible to anybody working on the system or responsible to manage it.

CHANGE DEFINITION

According to ITIL a change is "an action that results in a new status for one or more IT infrastructure Configuration Items (Cis)". Other IT procedure methods may provide similar explanations. The definition is quite large and may be subject to interpretation. In consequence, it is essential to clarify your understanding in this matter. In your change management procedures, and even sometimes in your incident management procedures, you should clearly define what is a change in your own IT environment. Otherwise people working in your environment will be continuously asking "should I create a change record for this action?".

For example: a critical server is frozen and needs to be rebooted. Will that be something that requires to apply the change management procedure? Some people may argue that since we're not doing any modifications on the server itself it should not require a change record. On the other hand, you don't want people to reboot servers without proper documentation, especially when dealing with scheduled maintenance. Following that logic some people will ask for a change to be registered, even in critical or emergency situations.

To solve this controversy some people have adopted the common-sense documenting approach. In other words: Everything that happens on a server must be documented, however there is no need to document actions twice. If a server has been rebooted you definitely need a record of that, but you can have this kind of actions documented in the incident database. There is really no need to have a record of the server reboot in the change management database as well.

Some people may voice their opinion stating that everything must be in the change management database. They will argue that in this way you don't really need to go elsewhere when looking for information on what has been done on the server. This approach, however interesting, doesn't reflect reality and can actually be misleading. The reason being is that even though you may be able to check all actions following this method, you will not have all the story behind the changes. You will always be forced to look at the incident database to understand the reason behind these actions. Therefore, when studying a server's (or any other equipment's) history, you need to consider the whole picture.

Be careful with people that like to take the easy way out and just ask to document everything as a change. Why will you adopt an interpretation that will restrict you from performing your duties and quickly correct a critical situation just because somebody's strict and narrow-minded opinion of a change? There is no medal award for those who have a stricter version of ITIL or any other recognized IT procedure method. In consequence, you should embrace an interpretation that suits your needs and provides the more benefits to your IT environment while conforming to a strict discipline.

Having said that, you still have to document any "new status" on your IT equipment on the change database. Although it may not be necessary to document all actions in there, you must clearly register any modifications that are performed, even in emergency situations. Let's say that you have a problem on a system that requires a configuration change; well, that is clearly a change. The server will not be the same after this action, so it will require a change record. Anybody looking at the change database will then be able to read when was the server's configuration modified and why was that done.

You should also consider automation when defining what is a change in your organization. Many things are now done automatically without intervention. This is a great improvement as machines get more intelligent. It will be absurd to ask the automation to create a change record and submit for approval for every automatic action taken to keep the servers up and running in a healthy manner. Although all actions must be documented, they are clearly not all considered as a change and will not require to follow the change management procedure.

In a virtual server environment, you will find that many servers will change from one host to another in order to keep a balanced environment among the host clusters. Some systems are configured to automatically increase memory, disk space or other resources on virtual servers as necessary. This is especially true in IT Cloud environments. There are many continuous actions being done in the environment; so how can we document all these changes?

Let's go back to an airline example: when a pilot sets-up the automatic pilot he has to do an analysis, take some decisions and give clearly instructions to his system. He will then set-up the desired altitude, speed, heading etc. Once he inputs all the necessary parameters he will turn the auto-pilot on and let it do its job. The plane will then automatically follow the pilot's instructions without any further intervention. It will move the ailerons, rudders and adjust the speed to conform to the pilot's initial command. It will not ask for intervention approval unless something gets out of the pre-established limits.

We can apply the same philosophy to our IT change environment when dealing with an IT Cloud system, virtual servers or any other kind of automation. A change record must be created and follow the normal approval procedure when we set-up the parameters and configuration. For example, we can limit how much memory or disk space can be assigned automatically by the system. We can also set-up the hosts CPU utilization limits so that virtual server charge can be properly distributed in the cluster. These limits are the ones we feel comfortable with; we can then let the automatic IT environment work correctly without intervention. In addition, we must set the alerts that will warn us whenever we're going out of the established bounds.

Now comes the discrimination question: while will you ask a human to submit a change for approval when you're not requesting that from a machine? It happened in one company I dealt with, they were requesting a change ticket for a technician to move a virtual server from one host to another. However, this is something that happens automatically, all the time. Are we doing this because we trust machines more than humans now?

There are several points to consider about this; first we have the documentation question. All actions done by the automation process are normally registered and can be easily consulted when needed. A human technician must document all his actions as well. It doesn't mean that he will always have to follow the change management and approvals procedure for all actions; but they must be registered somewhere.

Second, machine capacities are not the same as human capacities. In consequence, the limits and boundaries specified for actions taken by either author don't need to be exactly the same. A machine can do things quite faster compared to a technician that has to turn on his computer, login, take a cup of coffee and start typing stuff in his keyboard. In addition, all machine actions are pre-defined. It's like the instructions have already been typed in advance, so there is no chance of a type-o. In the other hand, some human changes can be standardized to imitate the automatic process. We'll discuss that in detail a little later.

But maybe the most important factor is the difference between the reason behind the action. All automatic actions are predefined and set-up in advance; when this happens then you do that. When a human has to intervene, he may not be falling into a pre-defined category; so, he will have to justify his actions. In other words, he needs to explain why he's doing the change. His intervention is kind of an exemption to the rule and must be explained and documented, in other words, a change record must be created.

On the other hand, we should not go crazy and ask for a change record for every single action on the system either. There will always going to be somebody who will justify the reason why a change ticket must be created with a statement like: "Can something go wrong with your action? What if something happens?". This is a protectionist attitude that goes beyond reality. In the real world things will happen, even if nobody touches the system. Besides, systems may crash or get in trouble without any intervention.

The truth is that there is always a risk. We normally try to minimize them or mitigate them, but the risk by itself should not be the main reason why we decide if a certain action should be considered a change. I agree, yes, risk is an important factor. But let's face it, there are no 100% risk free actions, there is always a risk. If the failure probability is low, then we should still study if a certain procedure should be considered like a change.

Actually, the risk along with the impact and other factors will determine what is called the risk importance. IT changes must be classified according to their importance. The reason behind it is that rules applying to these changes may be different according to how badly it is needed, how bad is the risk or what will be the impact. It's not the same to deal with a change on a server that is down and badly needs an intervention to recover compared to a server that needs a patch or when a minor modification must be performed. Some general rules will always apply to all changes; one of them being documentation. However, some approval methodology may differ depending on the circumstances.

Another factor to consider is the impact that a change may have. In other words, what will happen or who will be affected if the change goes wrong? If we are doing a change that may affect a five-people team working on a non-urgent project, then the change is relatively low impact. On the other hand, if the change may stop the whole business or considerably delay production then it will be considered a high impact change.

When planning an IT change, we should also consider its priority. Some changes need to be done right away, others can wait. All of these factors will help us determine the change type. We'll classify all changes in three different types: Emergency (also called urgent), Basic (with 3 different categories) and Standard.

EMERGENCY CHANGES

First let's talk about the emergency changes. This kind of IT change is required when an IT service is not functioning or has been degraded to a non-workable level. All of emergency changes must be associated to a high incident ticket. As mentioned earlier, not all actions will merit a change ticket; but they should definitely be done whenever the "status of the configuration item" is affected.

There should be no exception to this rule, this kind of ticket is only used during crisis situations. Emergency change tickets should not be used because somebody's project is late, because I absolutely need this or because the big boss demands it. There are other categories than can be used for these reasons. There is no use of having a change management procedure if anybody can just bypass the system for his own reasons.

One of the biggest arguments we here during a crisis is "just do whatever you need to do to bring it up again, we'll document later". The biggest problem with this logic is: Do what? Who's approving what you're doing? Whoever is working on the server must be able to document what he's doing. If he's trying commands to see which one works, he should be writing them down somewhere. If he doesn't, how will we ever know what works and what doesn't? If our star technician leaves the team and nothing gets documented, that knowledge will leave with him.

Most actions will need approval, even in a crisis situation. Therefore, they need to be communicated on the incident bridge. This way of working has a couple of benefits: the incident manager (or any other designated approver on the bridge) can evaluate the risk of any actions being taken; additionally, other technical people may object to the action mentioned on the bridge if they see a problem with it. Sometimes some actions taken to solve a certain problem can make it worse. So it's totally appropriate to take a couple of minutes to evaluate the activities.

Once all the appropriate actions have been identified, it's time to create the emergency change ticket. This should be a quick and easy process. RFC details should be reduced to a minimum in these specific cases; otherwise, if it takes more than a couple of minutes to create the ticket it will not be done and the process will just be ignored. If you insist in having all details, then the record may be created as soon as possible after the incident has been recovered. However, it should still be at least verbally approved on the bridge.

All tickets should be closed once the action is completed, even if we don't get the desired results. We cannot use the same ticket for other subsequent activities. One ticket may have several actions, as defined in the RFC. However, once the specified activities have been completed the ticket must be closed. Emergency closed tickets are an important tool for the RCA process.

STANDARD CHANGES

This kind of change is often forgotten during the change management process. Even though they are clearly defined in ITIL, Standard changes are commonly avoided in many IT environments. Nevertheless, they are an important tool to complete IT changes effectively while avoiding a good part of the bureaucracy that goes with change management process.

Standard changes are IT interventions that are routine. The procedure is well known, it has been executed numerous times and the risk is minimum. These changes are pre-approved by the CAB (Change Advisory Board). Therefore, there is no need to go through a complicated approval process and chase everybody down to execute them. So why will managers hesitate to use them?

One of the explanations that is usually given is the famous "what if?". In other words, we know we have done this several times without an incident, but what if something happens next time? For example, a virtual server may need to be moved from one host to another. Not only has this been manually done by system administrators dozens of times, but it also happens automatically quite regularly. Yes, there is a risk, there is always a risk. But again, we should be careful when we evaluate a standard change only because there is a risk. If we follow this philosophy, we will just never have Standard changes as there is not such a thing as a riskless action.

Some other people may argue: "but something did happen about a year ago (or some other period of time) when we performed this action!". The truth is that accidents do happen. However, if the incident was evaluated and mitigated, why shouldn't this action be considered a Standard change anymore? Using the same real life example, there was a problem while a technician moved a virtual server from one host to another. After investigation, it was determined that there was a problem with a specific firmware version being used. The problem was properly corrected, so there was no reason to remove this kind of actions from the Standard changes list.

There are a lot of benefits in using Standard changes. The CAB, change managers and other supervisors can have more time to concentrate on risky and difficult changes. The system administrators and technicians can better do their jobs as they can avoid the change management bureaucracy for actions that are well known and recognized. Standard changes are removed from the change report making it easier to read and identify possible conflicts. Even though changes are automatically approved, they are still recorded in the change management database and accessible to be consulted whenever is required.

The technical team or somebody who represent them should submit the list of Standard change candidates to the CAB. The CAB will then carefully verify them and call for a meeting when an explanation is necessary. All Standard changes will be pre-approved and documented in the Change management operations manual. Subsequently, Standard changes performed during a certain period of time must be audited to make sure that the proper procedures are being followed and the approved list should be revised periodically to determine if any adjustments should be done.

BASIC CHANGES

We'll now turn our attention to what is called Basic changes. These are the regular changes that need to be done from time to time in the IT environment. They should follow the normal approval and change management procedures. However, Basic changes can also be divided in different categories: Major, Significant and Minor. Changes belonging to different categories may be treated a little different.

You may read about each category in ITIL manual or any other IT procedures book. Nevertheless, you are responsible for customizing these procedures to your own needs. What is routine for somebody else may be critical for you and vice versa. The change management procedures should clearly indicate what kind of changes fall into which category. Some basic factors may be used for the initial classification. You should consider the risk (how sure you are that everything will go according to plan?), the impact (if something goes wrong, what will happen?), urgency (do you have enough time to plan for this?), outage time (any systems will need to go down during the intervention?) and how many people will participate on it (more people working on it will need a better coordination).

Some other people may classify their changes as minor, medium and high. The idea is to make a difference on how the regular changes will be treated. Regardless on their classification or type, they should all be submitted for approval. Normally they should be approved by the technical team performing it, the change manager (or change coordinator) and the CAB. They must be submitted using an RFC which will be discussed a little further on. They should also be listed and discussed during change meetings.

All changes must always have somebody who's responsible for them. This is the person who we can refer to and ask any questions about the actions to be performed during the change. He may or may not be part of the technical team. But he's responsible to coordinate with all the people who will participate in the change. He will represent the technical team at the change meeting and will be able to report if the change can be closed.

LEAD TIME

One of the main elements from the change classification will be the lead time necessary for the intervention. This is the determined time needed to plan and verify the change before the start time of the first documented action. The more complex, the riskier the change, the more time we need to better plan and have all the mitigation and emergency procedures in place for it.

This should never be treated like a bureaucratic procedure. If people view the lead time that way, they will just try to bypass it as much as possible. Lead time is important and should not be neglected. All changes need to be coordinated in order to avoid conflicts, make sure you have all the proper documentation for it and that the appropriate teams are involved. Then again, well documented changes should not be just waiting in line for the lead time to pass before the change manager, the CAB and the technical teams start looking into it. They should be discussed and supervised as soon as they are entered into the system.

However, as much as we explain the importance of proper lead time for every documented change, there will always going to be people that will want or need to be exempted. We should use the "expedite change" procedure for these cases. All exceptions should be validated by the change manager and the CAB before they proceed.

Some lead time exceptions may be correctly justified. For example: let's say we need an urgent intervention on a specific server. The server is not down (at least not yet) and service is OK. But if we don't proceed with a certain action there is a risk of an outage or service degradation. Therefore, there is certainly a proper sense of urgency to expedite the change. Even in these conditions, the risk, the teams involved and all other factors must be analyzed. Expediting a change doesn't mean throwing all important factors down the window because they are less important. What it really means is that the change will go over the top of the list and we'll be evaluated in high priority. Unverified changes, as important as they might be, could cause more harm than good if they are not properly planned.

On the other hand, some "expedite reasons" may be quite difficult to evaluate. Some people may want to push their changes to go earlier than what the normal procedure specifies because their project is late, some high-ranking manager demands it or even because somebody's vacation schedule is compromised. Many change managers will feel this is the time to put their foot down and deny these exceptions. We should be careful not to go that fast in doing so.

Somebody's project is late? Now, we should evaluate, how will this affect the company? Is our business liable to provide a certain service by a specific date? If so, as much as we want to complain about the project manager not planning his stuff correctly, we may still need to bypass the lead time and graciously give the green light for this change as we don't want to affect the business.

We may use the hospital's ER waiting list procedure as an example. Although hopefully nobody's life is at risk in our business, it may help us prioritize our work schedule. In many countries, you'll find a group of people waiting in the hospital emergency room to be treated. Some of them have the flu, some others may have a broken arm. They are entered in a list according to the importance of their illness. At a certain point in time, if somebody comes in with a heart attack, then any other patients with non-critical injuries will just have to wait.

A similar procedure will apply for change management. If a change needs to be done earlier, for whatever reason, it may be prioritized and put on top of the list. If it can be evaluated properly we may give it the chance to speed up the approval process when necessary. Then again, if something else comes up with a more urgent priority then the less critical change will have to wait in line. Having a good control on the lead time will help us go to the next step: scheduling the change.

SCHEDULING IT CHANGES

How to properly plan to the right date and time on a change? After considering the lead time to prepare for the change, we should determine when to perform it. The easy way out for many people is just to say something like: "you may do changes between midnight and 5 am". In reality, every system has its own individual needs that must be respected. We normally cannot arbitrarily just set-up a standard time for our whole IT environment. Therefore, each system should be properly documented with the appropriate time on which changes can normally be performed. In some cases, it will be at night or on weekends, in some other cases it may even be during the day at normal office hours.

Some people may object saying that their service can never be down, it needs to be working 24/7. The reality is that all systems need to be maintained and undergo technical intervention from time to time. If a certain system or application needs to be available all of the time, then it should have proper redundancy to accommodate maintenance. We can use clustering, virtualization or any other method to keep the system running while performing the necessary technical work on it. It's just in a utopic world that there will never be absolutely no intervention on a certain system.

Another important factor to consider is the allocated time for the change. This can be trickier with today's change management philosophy. Many people will suggest using the worst case scenario to plan for the appropriate time frame. In my experience, I received changes asking for 36 hours for a server reboot. When I asked why they required so much time for a simple 15-minute operation they said: "Well, what if the server doesn't come back after the reboot? And if there is a hardware failure we may need at least 24 hours after that to receive the replacement part. Plus, the time for the technicians to come in and repair the machine will take us additional time!".

As absurd as it might seem, these technicians that over cushioned their change had an important reason to do so: "We penalize and punish people that don't respect their time frame". Some people had put their job security at risk because they had gone over their allocated change time by about 30 minutes. They had an unexpected event that delayed their change and then they had to go through several meetings to explain their actions and justify the additional time. Moreover, an unreasonable manager wanted to have somebody fired to make the client happy.

This example demonstrates the importance of being reasonable while calculating the allocated time for a change. We want to have some cushion time available in case something happens. But we cannot predict and allocate time in case we have an alien invasion while working on a change. We document the risk factors, but then ask for a reasonable time with a sound cushion (or realistic extra time) for our change. On the management side, we must examine failed changes and verify if any additional documentation or procedures may have prevented the problem.

But the scheduling process is not finished yet, we're still missing the most difficult part: coordinating all the changes. The change manager or change coordinator must validate all changes submitted to him, list them and verify for any possible conflicts. He should normally organize a change meeting with the purpose of having all parties work together and make sure that there are no conflicts with the proposed schedule. The change meeting should be done as frequent as required. Some organisations will do them weekly, some others even daily. It will depend on the particular needs of the organisation in question.

If often happens that a change must be re-scheduled. There may be resource conflicts, change conflicts or technical difficulties that will push to postpone or advance a change. Every change must be approved with a start and an end time. If those times have to be adjusted, then the whole change needs to be re-evaluated and approved. The change coordinator must verify for any new conflicts and make sure that everybody is aware of the new time frame.

CREATING AN RFC

All IT changes must be submitted by creating an RFC (Request For Change) or any other similar document approved by the CAB. All RFCs or change requests should be kept in the change management database and be used to document any actions to be performed during the intervention. By accurately documenting every IT change we can avoid misunderstandings and support good communication between all concerned teams. In this chapter, all reference to the change ticket documentation will be called an RFC, however you may use other names or title is you're not following ITIL.

The first rule on an RFC is that it must contain all the documentation necessary for the change. It means that it cannot just have references to find some information somewhere else; everything should be in there. Nobody should need to look elsewhere to understand the change. Also, the person responsible for the change should own the RFC. A step by step procedure of all actions to be performed must be contained in the ticket. It should normally have a ticket number or some other unique reference so anybody can point to the right RFC when it's been discussed.

The intended start time and end times should be clearly indicated in the RFC. Additionally, we can have outage times, identified time frame, etc.; but they are not mandatory. It should also contain the list of approvers for the change. It should be unambiguous, clear to anybody reading the ticket to know and understand if it's approved or not, and who's missing to approve it. The ticket owner should be responsible to make sure that the ticket is ready and approved before the start date.

Another important part of the change documentation is the back-out and mitigation plans. Preferably we'll mark a back-out procedure, how to undo whatever was done. For example, if the change is to modify a certain code for an application, the back-out procedure should describe any actions needed to put the old code back. Just as the change documentation, it should not contain any references outside the RFC; all information should be in the same ticket.

Unfortunately, it's not always possible to back-out of a change. Some people may argue that there is always a way to do that and it must be documented. That statement is right, but not exact. Back-out means we will erase what we just did and go back to the previous state. In some cases, that's impossible! Consider a server reboot for example, you can't undo a reboot. Once it's done, it's over. You cannot tell the server to ignore the last reboot. That doesn't mean there's no need to have a plan B, but in this case, it should be call a mitigation plan, not a back-out plan.

A mitigation plan will describe the procedure to follow whenever we experience a problem on which we cannot back-out. In the reboot example above, a mitigation plan may comprehend a procedure to call for on-site support, hardware support or have console access to troubleshoot the situation. It's true that we cannot predict every single issue that may occur, but we should plan a way out in case something happens, especially for the most obvious problems.

SERVER PATCHING

Server patching is the common name we give to the activity on which we put the operating system up-to-date. It's normally done by downloading the latest package published by the vendor and applying it to the servers. Sadly, I've seen that many companies have very bad planning for it, even if it's done regularly.

Patching should be looked at as a certainty; it will happen and you cannot avoid it. At least, you shouldn't. First of all, the security of your servers may be compromised. Even if you have the greatest firewall and network protection in place, it's always good practice to have all the proper security patches installed. With so many computer viruses, Trojan horses and other malware going around, you cannot take chances and avoid patching.

Additionally, it may help solve issues on your system. The first thing the vendor will ask when you call them to report an issue is: "Is your server up to the latest patch level?". Even if you haven't encountered any problems with your current configuration, it doesn't mean you won't in the near future. Better avoid the headaches and have all your systems up to date.

Some people may object to that by saying that the latest patch crashed their system. Yes, unfortunately it could happen. Although, it's not common, there is a possibility that applying a patch will make things worse. But that doesn't mean you should stop patching; what that means is that you should be testing your critical systems to make sure the patch will not adversely affect you. In some cases, this could have already been done for you. If your using a vendor's application, you may always contact them and confirm if their application works correctly with the latest patch.

Another obstacle for regular patching is system validation. In some special cases, IT managers are responsible for doing a full test on their applications every time there is a change in the system, which includes patching. This can be a lot of work in some cases. But just think in the amount of work it will take if your system gets hacked or starts crashing without notice. As much work as validating can be, you cannot use it as an excuse to avoid patching. You can however try to customize your validation procedures to fit your needs and speed up the process.

I feel quite disappointed to witness people running around getting approvals so they can apply the latest patch. It should be a well-planned and scheduled activity. Nobody should be surprised and ask, "why do you need to reboot my system?". It's normal, and it needs to happen regularly. As such, it should be scheduled in advance. Even if we don't know the exact patch to be installed or who will be working on it we can have some predetermined dates for this activity.

For those using Windows servers, you should be aware the Microsoft will normally publish a new patch every month. It is wise to follow the vendor's advice; even if you think your server is properly secured on your network. You may schedule to apply your patches on the first or second weekend every month. Or maybe even distribute your systems to be patch on different weeks if you have an important number of servers.

Normally you don't need to patch Unix servers every month. Nevertheless, you should be following the vendor's advice as well. Many companies regularly patch their Unix servers every 6 months.

Besides the operating system, you should also be aware to patch the software running on your servers. This may include Oracle, SQL or some other database software. It may also include any other special software you may have installed. As mentioned above, it's prudent to test any patches on your software before applying them to your production environment.

You must add your patching level information to your server inventory in order to make sure that all your systems are properly patched to the latest level. This document will quickly help you identify any missing patches in your environment. It may also be a powerful tool when auditing your procedures. As a matter a fact, you should always keep your patching procedures to be re-used during the next patching cycle. Some systems actually require considerable amount of work to plan for it and all of that valuable information may be lost if you don't use a patching database or some other kind of patching knowledge collector.

WHEN SOMETHING GOES WRONG

We have the change correctly documented, we have the proper teams and we went through the planning meetings. But what if, at the last minute, something unexpectedly happens? The response should be equivalent to the crisis level. It is normal that we may find some stuff out of place, or that the equipment's behaviour is not exactly what we thought it will be. However, we must evaluate the situation and apply the proper mitigation procedure.

Can we continue to follow the same procedure and arrive to the same results? If that's the case, we should just continue and document our findings for later study. If there's no outage and we don't really need to change our plan, then there's really no reason to panic. Any new findings can be discussed in a later time, after the change has been closed.

Now what if our findings indicate that we have to do a serious deviation from the approved procedures? At that point, the change manager should evaluate the situation with the change owner. They must verify the risks and the impact of applying the new suggested solution. Depending on the change management procedure being followed, it may also be quite appropriate to notify the CAB and get a new approval for the change. This will also apply if we significantly need to extend the change window. It may happen that some unexpected results may have delayed the change. Having a second approval should determine if we can proceed, need to cut the change short or even back-out the change.

But let's face it, quite often when things go wrong, they really go wrong. So what if our change is causing an unexpected outage? At that point the best practice is to invoke the incident management procedure. It's better to have an incident manager evaluate the situation and have the appropriate people on the bridge to recover the incident as well as to take important decisions about the change in question. At this point in may be more important to recover from the incident rather than it is to continue and finish the scheduled change.

Regardless of the severity of the incident or the situation that happened, it should be taken as a learning opportunity. After the problem is solved, we should document what happened and how the issue was solved. The incident management and/or the change management procedure should be updated with the new information. If new procedures need to be applied they must be communicated to the concerned teams.

CLOSING THE CHANGE

It looks like a simple question when we ask when to close the change, doesn't it? The simple answer should be to close the change ticket when the change is completed. It may not be that easy in real life and many technicians and system administrators struggle with their management teams about this.

The reason behind this dilemma, is that after a change is completed we may still experience good or bad outcomes from it. Some people like to wait a few days to make sure the change didn't cause any unexpected trouble. It's true that it may sometimes take a couple of hours, and even a couple of days, to be completely sure that a change was successful. In certain situations, we may uncover a problem that was caused by a change done months ago! So, how long do we really have to wait to close the change record?

The answer may be surprizing, but it's back to the logical, common sense answer: "as soon as the change it completed". The biggest problem with this is not the logic behind this, but the change management systems in use by many organisations. Some systems will not allow any updates once the change is closed. This is a big mistake in my view! The change records need to be updated to correctly describe the history of what happened. Changing this procedure is not easy; sometimes failed changes are accounted for penalties and evaluations. Following this, the applications used for change management will not allow any updates because of the politics and finance issues behind this.

The opposite is sometimes true, a change may be marked as failed, and later on we discover that the incident had nothing to do with a certain action. We may find out later about the real reason behind what was supposed to be a failed change and update the documentation created for it.

No matter what people say, having accurate records must still be the rule. You may put time limits to penalties and evaluations as you need to do the cut somewhere. But you cannot maintain a history of a change following a certain procedure as successful when it really wasn't. People should be able to trust the information in the change management database and understand what happened when a certain change was applied. You may work with whatever accounting method you want to use, but keep your history truthful. The change should then be closed with the correct status soon after the change was completed. Any new information after that should be added to the change record as appropriate.

CHAPTER 8

Service Desk

"Oh, those poor ignorant people asking for help!" What a bad way to refer to people calling the Service Desk. How many jokes do you know about people calling and asking for help while mentioning something that is quite obvious for somebody who has more IT experience? Unfortunately, we have been mocking the users for too long. Even worse, the same attitude has driven how users are often treated when the call the Service Desk for help. We build our procedures in such a way that we can get rid of them quickly. We treat them as IT illiterates, fools or even as a nuisance. How dare they call asking for help?

Enough! These people need help and should be respected. The truth is that nobody knows it all, we all need help in a certain point of time. Many of these callers may not know much about computers, but they may be geniuses in their own field.

The purpose of the Service Desk (also known as the Help Desk) is to provide assistance for people who may have some trouble using a certain system or application. They are our clients in the IT world and must not be ignored. Some managers had made big cuts in these departments, arguing that the users will find their way somehow. They are right in that point, people will find their way, they are not stupid. But in order to do so, they might ask a colleague, call somebody or waste valuable time trying to resolve their issues while making the business departments more inefficient. What initially looked as good savings can end up costing more in the long run.

INITIATING THE CALL

"Hello, how can I serve you?" We probably all know the traditional salutation when we call for help on a certain subject. We expect to hear a friendly voice at the other end from somebody ready to assist us. Sadly, many people have the senseless idea that they can scream, shout and sometimes even insult the person on the service desk while they remain calm and gentle. Honestly, some people working at the Service Desk should get a medal! As much as we want to assist people experiencing problems, we should never allow rudeness.

On the other end, the traditional way to start a call is by picking up the phone. We then often hear a message reminding us how important our call is for them while we keep on waiting. Some people may have to stay on the phone waiting list for nearly an hour before being able to talk to someone. In the IT world, this is totally unacceptable. In a certain company the user often called the technician directly, and then followed the procedure and created an incident ticket with the Service Desk, just to follow the process.

In this case the Service Desk procedure was totally useless. They were standing there just like glorified secretaries taking the information from the clients and creating the ticket. That is not what the Service Desk stands for and should not be used in that way.

So how long should we keep people waiting? The answer is that you get what you pay for. How long are you willing to wait for your service to be down before it's restored? If it takes a client about an hour to contact the Service Desk, it means that the service will be interrupted at a minimum for the same amount of time. There are other ways to get things repaired, but the Service Desk will normally be one of the main methods used by users to signal that something is wrong with the system. While the phone remains the traditional and sometimes preferred contacting method by most users, there are plenty of other ways to contact the Service Desk.

Another popular method is using a website. It has the advantage of saving Service Desk people from typing from what they hear on the phone and making mistakes. In this case, the client itself is responsible to document his ticket. His name will be clearly spelled and it may avoid a lot of confusion as he will normally describe the problem on the ticket. The disadvantage is that the client needs to have a computer and an internet connection to login to the Service Desk website, and problems with his computer and internet is often the reason why he's calling in the first place, especially in IT. So as much as you want to push the clients to use websites, you still need an alternate contacting method.

Some people are using text messages or some other new communication methods to reach the Service Desk. These may be good tools to consider, they all have their advantages and handicaps. The importance is to be able to contact the Service Desk department and provide enough accurate information for them to evaluate the incident and record it in the database. However, the more important step they should take is to actually provide the service their meant to offer and make sure the issue is solved.

SOLVING THE PROBLEM

When people contact the Service Desk they have one thing in mind, they want their issue to be solved. They definitely appreciate the kind salutations, friendly conversation tone and all the nice tools to communicate their problem; but all of that is totally useless unless the trouble is solved to their satisfaction. Sadly, the most important objective of this department is often neglected. We can track calls, print out incredible precise reports and do all kind of magic with the data. But how good are we in getting the problem solved? We really need to focus in the main objective of this department.

Solving issues could be easier said than done. Nevertheless, the logic behind it is not that complicated. To simplify the response, we'll normally divide it in two or three levels. The first level will comprehend the general knowledge questions. It will include things like: "How do I print this document? How do I access this system? Or even, what is a spreadsheet?". There are several ways to provide the right answers to the callers. In certain organisations, they have a level 1 technical response staff, in some others, the Service Desk itself will act as a first response team. Either solution has its pros and cons and must be evaluated according to your own needs.

When we use a first response staff, the Service Desk's behaviour will be simple. They will just answer the call, record the information and create the ticket. The ticket is then transferred to the appropriate team who will solve the issue either by providing the correct response or taking the appropriate action. In some cases, the ticket can also be directly created by the user by using other contact methods like described earlier. This process has the advantage that you don't need a highly trained Service Desk staff as they are just capturing the information. It may be useful in situations when you have a large number of calls and you need to treat them quickly in order to provide a good service to the users. The disadvantage is that the response is normally delayed as there will be a time lapse between the moment the call was issued and the time on which the first response staff will work on it.

The other option is to have the Service Desk provide first response service. Using this option, the staff will have some general knowledge and will be able to respond to general questions asked by the callers. They will also be responsible for taking the first look at tickets created using a web site or any other method. The advantage on this method is that the callers will have a quick response for their questions once they get in touch with someone. The problem is that the calls will usually take longer in this case. This may translate a longer wait time on the line for callers which normally sets off a high level of frustration. In most cases the best solution is to have a good balance between both options. In other words, you equip the Service Desk staff with enough knowledge and scripts so they can give a quick assistance on the more general questions, however you use the first response staff if the issue may take a little more than a few minutes to resolve.

Scripts are useful in these circumstances. Each script will contain a situation and how to respond to it. For example: A user calls to complain that he can't access a certain application. The Service Desk staff can use a script to quickly troubleshoot and guide the user to the solution he's looking for. It will be prepared by a knowledgeable technician and may contain questions like: "Can you access the web page? Are you able to login? Can you see the welcome page?". Each negative answer will drive to a possible solution that will be recommended to the caller.

If it looks like the issue is out of the normal troubleshooting knowledge, it may need to be referred to the next support level. This level should be staffed with more expert technicians that can evaluate and study the problem more deeply so they can provide the proper solution. Whoever is managing the call at this point should be able to transfer the ticket to the appropriate department or assignee. Somewhere there is someone with the right answer to the question. The correct procedure has to be in place in order to transfer him the information so he can respond in a timely manner. Regardless of the method or procedure being used, it must include a way to get back to the original caller and close the call. We cannot assume the issue was solved because the technician believes the situation is OK now, it must be confirmed with the original caller.

MANAGING HIGH SEVERITY SITUATIONS

It's bound to happen; at a certain point of time the Service Desk will receive a call or a ticket that will trigger the red alerts. Users may be reporting a system outage, slow-downs or a major event that is seriously affecting the business. The right procedures need to be in place so that emergency situations can be treated quickly and effectively. The last thing we want is having people waiting an hour on line to report an incident with dire consequences.

High severity or high impacting situations may not be as easy to detect for the Service Desk staff. A user may report he can't access a certain system, this may be because he lost access himself, or it could be because the system is down. Once again, having a good script and asking the proper questions is crucial to detect any major anomalies. There are certain factors that can be considered in advance and can guide the Service Desk staff to recognize if they are facing a major event or service outage.

Another important tool is to have good communication and collaboration among the Service Desk staff. Let's use the same example from a user complaining that he cannot access a certain system. If the person taking the call is aware that his colleagues are getting the same complains it will help him quickly recognize that this is not an isolated incident. Once we're able to do this we can save precious time by immediately invoking the incident management procedure instead of frustrating the caller by prolonging our questionnaire and asking him to take useless actions like rebooting his computer. The Service Desk should be able to rapidly scan the incident database when they get a call to verify if other similar incidents have been logged.

Once a major incident has been recognized and the proper procedure has been followed, we need to have a way to communicate to everybody that the issue has already been reported and that we're working on it. Not only must the Service Desk staff need to be aware of it, but it also needs to be properly communicated to the whole IT community. Some organisations use information boards so that Service Desk staff can quickly be informed about any important issues when they get calls from the users. Outage information can also be posted in the Service Desk web site. Additionally, we can include voice message recordings about known problems in the Service Desk phone system so that users calling will be realize that the issue they're reporting is currently being worked on.

A special procedure must be used to report high severity incidents. It cannot just be a ticket in a database waiting to be noticed by someone. The Service Desk staff should immediately contact the incident resolver. The resolver should be the person on the right position to understand what is happening and perform the corrective actions to solve it. It most cases, the resolver should be identified in the system's incident management procedure. In addition, there should be a "warm transfer" between the person reporting the issue and incident resolver. In other words, there should be a direct communication in which the person reporting the incident will provide all useful details about the situation.

Shouldn't the Service Desk call the Incident Manager to report the issue? Yes, it could be quite useful if they do, depending on the type of problem being worked on and your own practice in the IT environment. Some organisations will have the Service Desk call the Incident Manager, and some other will have the incident resolver or the Service Manager do that. Nevertheless, the incident resolver should always be the first person to be contacted. If they call the Incident Manager first, and have him call the incident resolver you'll be wasting some valuable time. By putting your calls in the right order, the resolver may start working on the incident while the Incident Manager prepares his documentation, opens the bridge (or discussion conference call) and calls everybody.

EVALUATING THE SERVICE DESK

In any organisation you need to know how well your procedures are working. In the case of the Service Desk it's particularly important to define the right KPIs to determine if you are meeting your main goal: to provide professional assistance to your users in a timely manner.

In order to achieve this, you cannot just count the number of calls and how fast they were closed. That may give you the wrong information. If you're only evaluating your team on those parameters, you'll end up with a staff that might be too quick to close unresolved tickets just to make the numbers look right. Those factors are important, but they should be compiled with other components to provide a better picture on the service being provided. Other important things to consider may be: How many calls were actually resolved by the Service Desk? How many calls were transferred to the next support level? How much time elapsed between transfers? How quick were high severity incidents identified and transferred? Etc.

Nevertheless, numbers will not be able to tell the whole story, so you may need to resort to surveys or other client satisfaction evaluation methods if you want to make sure your Service Desk is performing according to expectations. The problem with this solution is pushing people to take the time to respond to the surveys. This is especially true when things are going well; people will often take some time to complain when they find problems but will not be that eager to participate in surveys just to say that everything went OK.

Some companies like to make long and complicated surveys to gather a lot of information, but the end result is that nobody will complete them. When using surveys, they need to be short, precise and easy to respond. Otherwise you may need to provide incentives so that people participate on them; even a small incentive will do. Following human nature, people will go out of their way to get to the gas station that is offering a one cent discount for every gallon/litre of gas, so small incentives do work! You may offer a chance to participate on a giveaway, or even a 10% discount on coffee will do. Yes, there is a cost; but you cannot always expect people to grant you their time for free. If you assume that everybody will willingly participate on your long and complicated surveys, you'll only get a small chunk of the client population and you'll still miss the whole picture.

Understanding your performance is great knowledge, but it's absolutely useless if you don't do anything about it. Procedures need to be revised and corrected, ticketing systems need to be adjusted, staff needs to be trained. The information you'll gather may certainly help you provide the business case required when additional investment is necessary. In addition, don't forget to performance information from your staff, they may have some valuable ideas that may help you improve the service.

ABOUT THE AUTHOR

Antonio Narro has been working as an IT Service Manager for about 10 years. He has experienced different managing methods used by various companies. He has been able to recognize some typical mistakes often made in today's environment. He started working in IT as a system administrator and programmer before working in service management. After 28 years working with different companies and in different environments has seen the evolution in the methods used to manage computer systems.

When Antonio started working in IT there was a lack of good procedures and strict disciplines. Some small businesses were managed by an expert guru that was used to take all the IT decisions for the company. Gladly, the IT environment has evolved and he has seen the transition to the new structured IT environment specially used in big corporations. While he appreciates well documented procedures and sees the importance of appropriate documentation he normally recommends to always use common sense when applying them.